Ion Exchange for Drinking Water Treatment

Ion Exchange for Drinking Water Treatment

JOHN CIVARDI

American Water Works Association

Disclaimer

The authors, contributors, editors, and publisher do not assume responsibility for the validity of the content or any consequences of their use. In no event will AWWA be liable for direct, indirect, special, incidental, or consequential damages arising out of the use of information presented in this book. In particular, AWWA will not be responsible for any costs, including, but not limited to, those incurred as a result of lost revenue. In no event shall AWWA's liability exceed the amount paid for the purchase of this book.

Editorial Manager – Book Products: Melissa Valentine
Copy Editor: Suzanne Snyder
Product Manager: Tony Petrites
Production Manager: Gillian Wink
Compositor: Innodata
Cover Design: Michael Labruyere

Library of Congress Cataloging-in-Publication Data

Names: Civardi, John, author.
Title: Ion exchange / by John Civardi
Other titles: Ion exchange for drinking water treatment
Description: First edition | Denver, CO : American Water Works Association, [2021] | Includes bibliographical references and index. |
Summary: "This handbook provides information regarding the selection and design of ion exchange processes for water treatment. Additionally, it aims to enhance the understanding of the "package type systems" that have become so common throughout the industry"-- Provided by publisher.
Identifiers: LCCN 2021018980 | ISBN 9781647170462 (paperback)
Subjects: LCSH: Water--Purification--Ion exchange process. | Drinking water. | Water quality.
Classification: LCC TD468 .C58 2021 | DDC 628.1/62--dc23 LC record available at https://lccn.loc.gov/2021018980

Printed in the United States of America

American Water Works Association
6666 West Quincy Avenue
Denver, CO 80235-3098

ISBN 978-1-64717-046-2
ISBN, electronic 978-1-61300-585-9

CONTENTS

FOREWORD

I make the pollution; you clean it up. This statement was made by the finest engineer I have ever known: Frank Peter Civardi, my father. You see, my father was a chemical engineer from the 1950s through the 1980s, and his career was dedicated to the development of chemical products that have contributed to our modern world. This book is about "cleaning up" pollution in drinking water using ion exchange treatment. My goal is to share with you a bit about my journey in the development of this book. This book is the fulfillment of my goal to help the drinking water community in the practical implementation of ion exchange to reduce the harmful effects of specific chemical contaminants found in our drinking water. My hope is that knowledge of my journey can inspire you in the pursuit of your goals and dreams with regard to the production of safe drinking water.

In 1977, I was 12 years old, and I was beginning to notice degradation of the environment in my neighborhood. I spoke to my father, and he suggested I write a letter to the president of the United States, Jimmy Carter, describing my observations. I followed my father's advice, and I received a card back from the office of the President. My father was so proud of this card, he hung it in his home office for many years. This event was my first entry into what has become environmental science and engineering, and it taught me that words matter and that my ideas may be worth listening to. A few years later, my father, who was born and raised in Italy, decided he wanted to take a family vacation to visit some of America's national parks, including the Grand Canyon. He packed up the car and we drove from our home in New Jersey to "see America." It was on this trip that I discovered the immensity, grandeur, and beauty of nature, and I knew in my heart that I wanted to somehow be a part of protecting the environment.

I was very close to my father, and in some ways, I believe I was one of my father's engineering projects. When I was in high school, he encouraged me to focus my studies on math and science, and when it came time to select a college, he asked me to consider only engineering schools. During my college visits, he would speak with professors, and it was our "collective" opinion that Stevens Institute of Technology was the best place for me. When I entered Stevens in 1982, there were about 1,500 students in the undergraduate program, and approximately 80 percent were engineering majors with the remainder majoring in one of the sciences.

For the first two years, all engineering majors took the same core classes, and dropping a class resulted in being placed on academic probation. The curriculum trained the students to be rigorous in their pursuit of excellence. In my junior year, all engineering students selected their specific discipline, and much to my father's dismay, I selected civil engineering. You see, environmental engineering was relatively new at that time, and Stevens only offered a master's degree in environmental engineering. My father was not pleased. The 1950s through the 1980s was an era of explosive growth in the chemical engineering profession, and New Jersey was the epicenter of this growth. Companies such as American Cyanamid had their corporate headquarters in New Jersey, and many major chemical companies had large operations in New Jersey. Many of the products—including the keyboard I typed this book on—were the result of the efforts of chemical engineers, and my father wanted me to be a part of the profession that he loved so much.

My father allowed me to take up civil engineering, and in my junior year, I had my first hydraulics course. That's when I realized that there could be a connection between engineering and my desire to improve the environment. I was awestruck by the fact that engineers could accurately predict the physical behavior of water in the environment. But I wanted more, and Stevens allowed undergraduate students to take classes in their master's program when they were still undergraduates. The master's courses were offered at night, and I was able continue with my undergraduate classes in civil engineering. I decided to take night classes in environmental chemistry and environmental biology. A whole new world was opened to me, and I was beginning to realize not only that engineers could use science to "make" things but that the things we make can have adverse consequences. More importantly, I discovered that environmental engineers could address the legacy of harm that had been caused (albeit in many instances unknowingly) by the industrial revolution.

When I graduated in 1986, the era of environmental regulation was in a growth phase. I witnessed the promulgation of the surface water treatment rule and regulations of volatile organic compounds. The legacy of pollution from the chemical industry was beginning to be understood due in part to

- our ability to measure these compounds at the part per billion level, and
- the ability of the public health community to demonstrate the health impacts associated with consuming water at seemingly low levels.

During my career, my father remained a source of guidance, and he would inquire about my projects. In the early 1990s, I was the project engineer for a new plant in Hyde Park, New York. There was a dedication for the plant, which families were invited to attend. I took my wife and young son and my mother and my father. My father was so proud of me, and he could not believe that his son had some small part in this complex treatment facility. As he had said, "I make the pollution; you clean it up."

On April 11, 2001, at the age of 70, my dad passed away, and my mom asked me to go through his technical books. It was at her home that I discovered a book entitled *Ion Exchange Technology* by Nachod and Schubert, dated 1956. The book is over 600 pages and covers a broad range of ion exchange topics. My father never mentioned this book to me. You see, as my father advanced in his career, he was actively involved in research and development; however, before I was born he was a process engineer and I suspect he must have felt some kind of attachment to this book from his early days as a chemical engineer.

In 2001, I was not working on any ion exchange projects, but I decided to keep this book, perhaps as a reminder of my father. In drinking water, ion exchange has historically been limited to softening, and there is not an AWWA standard for ion exchange media. Over the past 10 years, however, there has been renewed interest in the use of ion exchange media within the drinking water community. I have had the opportunity to work on several projects involving ion exchange. In working on these projects, I found there was not a single publication dedicated to assisting utilities, engineers, and regulators in the evaluation and conceptual design of ion exchange systems for drinking water. My father and his book from 1956 became the inspiration for this handbook.

All journeys and projects require a dedicated team, and I want to thank the following individuals and organizations for their efforts, without which this handbook would not have been possible:

- Tony Petrites and the editorial staff of the AWWA for their patience and dedication
- Barb Martin and Corinne Bertoia at AWWA for their technical review
- Francis Boodo, Ted Begg, and Cathy Swanson of the Purolite Corporation
- Rick Zahnow of Water Remediation Technologies
- Ellen Gaby and Geordan Stroud of Ixom
- Michael Pickel, Director of Compliance & Regulatory Affairs of Horsham Water and Sewer Authority, who provided the PFAS Case Study
- The following staff members from Mott MacDonald:
  - Maria Borman, who provided input on the Piloting chapter
  - Keith Cartnick, who provided input on the Simultaneous Compliance chapter
  - Kuteiba Hussein, who provided input on the Water Reuse chapter
  - Kenny McGibbon, who coauthored the Nitrate and Water Reuse chapters
  - Stephanie Douglas, who coauthored the Perchlorate chapter
  - Margie Gray, who coauthored the PFAS chapter
  - Meredith Raetz, who was responsible for compiling all the chapters and organizing the large amount of references
  - Mott MacDonald (Jason Marie and others), who provided me with the time and resources to complete this project

And, finally, I want to thank my father for having faith in allowing his son to pursue his dreams. Remember: "I make the pollution; you clean it up."

ABOUT THE AUTHOR

John Civardi is Mott MacDonald's Global Practice Leader for Water and Wastewater Treatment. He has 35 years of experience in the evaluation and design of drinking water systems. He is a member of AWWA's Ion Exchange Standard Committee and is chair of AWWA's Membrane Design Subcommittee. He was the coauthor of AWWA's Iron and Manganese Handbook, 2nd edition. John is a licensed professional engineer in New Jersey. He received his BE and ME from the Stevens Institute of Technology, Hoboken, New Jersey.

CHAPTER 1
Introduction

1.1 Overview

The purpose of this handbook is to provide drinking water and water reuse professionals with information that can be used to answer the question,

> *"If my system has a contaminant approaching or above the maximum contaminant level, is ion exchange a feasible technology and if so, what would be the basic characteristics of the ion exchange system?"*

This handbook provides information regarding the selection and design of ion exchange processes for water treatment. Additionally, it aims to enhance the understanding of the "package-type systems" that have become so common throughout the industry. To accomplish this goal, information on the resin composition along with a description of the operation of ion exchange systems is described. A fundamental component of this handbook is the use of design examples and calculations provided at the end of many of the chapters. The goal of these examples is to provide insight into the sizing of the process components and enable the user to assess potential impacts on changes to design criteria. Along with these design examples, several case studies are included.

This handbook describes the contaminants often found in drinking water that can be efficiently removed with ion exchange, including emerging contaminants such as per- and polyfluoroalkyl substances (PFAS). A discussion of post-treatment topics is provided, including

- simultaneous compliance,
- impacts on distribution system water quality, and
- residuals.

In many instances, site-specific testing is needed to assess the feasibility of ion exchange, and guidance is provided regarding the performance of bench, pilot, and demonstration studies.

The ion exchange handbook will be a valuable resource for

- Water system owners (municipal, industrial, federal, and private), operators, engineers, managers, laboratory staff, maintenance staff, and procurement managers. These individuals are responsible for planning and implementing treatment solutions.
- Engineers. Consulting engineers are typically tasked with planning and designing ion exchange systems as well as managing pilot systems.
- Regulatory agencies (federal, state, and local). It is important for these agencies to have a basic understanding of treatment technologies capabilities and limits, along with applications where ion exchange is feasible. This handbook also includes regulations from the United States as well as international regulations/guidance.

This handbook will be most useful at the initial planning stage of a project, as well as during the 30 percent design phase to establish if and how ion exchange should be incorporated into the treatment system. It enables the user to select the unit processes and establish the design criteria for the ion exchange system. It is not intended as a detailed design manual or comprehensive guide for ion exchange in water treatment. Ion exchange systems contain multiple process, mechanical, and electrical components as well as chemical feed systems. This handbook does not cover the design of these components. The goal is to provide guidance on the uses and implementation of ion exchange systems in drinking water treatment. This handbook also provides information related to start-up and commissioning of the ion exchange system upon completion of the design.

1.2 Questions That Can Be Answered

This handbook can be used to address questions such as

- What is ion exchange, and what are the types of ion exchange that can be used in drinking water and water reuse applications?
- What drinking water contaminants can be removed by ion exchange?
- To what levels can ion exchange remove particular contaminants?
- Will ion exchange be feasible for my site/facility?
- What water quality parameters are needed to evaluate the feasibility of ion exchange for my facility?
- What equipment is needed for my ion exchange system?
- What are quantities and dimensions of the ion exchange system for my facility?
- What are the flowrates, volumes, and composition of the residuals generated from the ion exchange system?

- How are ion exchange systems controlled, and what instruments can be used to monitor the performance of ion exchange systems?
- When is bench, pilot, or demonstration testing needed, and what is the framework for performing these tests?
- What are key components and considerations of a pilot system?
- What are the potential impacts of simultaneous compliance issues?

1.3 Ion Exchange for Drinking Water Approach and Organization of the Handbook

This book is organized using a "contaminant-based approach." An overview of the ion exchange process is provided in Chapter 2, and then individual chapters are provided on the major drinking water contaminants that can be removed using ion exchange. Each of these chapters is a "stand-alone" chapter that can be used in conjunction with Chapter 2. Following the contaminant-specific chapters are chapters regarding topics ancillary to ion exchange systems such as pilot testing and commissioning. The approach used in this handbook is described in Figure 1-1.

The following presents an overview of each chapter:

Chapter 1—Introduction: This chapter explains the purpose of the handbook and the water-quality parameters that can be addressed with ion exchange. It also provides a description of the topics that are covered in the entire handbook.

Chapter 2—Fundamentals of Ion Exchange: This chapter describes the chemistry of ion exchange, the types of resins, and an overview of the equipment components. There is no AWWA Standard on resins, although AWWA is intending to develop a new standard. For drinking water applications, the resin should be approved by the appropriate regulatory body and agencies such as NSF International, state agencies, and international organizations. There are several suppliers of resins each with several product offerings. This book does not provide a listing of suppliers and resin offerings. Where resin suppliers and products are described, this information is primarily for illustrative purposes.

Figure 1-1 Ion exchange handbook organization approach

Chapters 3 through 11 describe ion exchange for the removal of

- hardness,
- total organic carbon (TOC),
- radium and uranium,
- arsenic,
- nitrate,
- perchlorate,
- chromium,
- PFAS, and
- water reuse.

Each of these chapters discusses the regulatory issues associated with the contaminant, the source water quality parameters that can impact the resin, an overview of bench and pilot testing requirements, design guidelines, and a design example.

Chapter 11—Water Reuse Applications: This chapter discusses the role of ion exchange in water reuse facilities. The unique water quality, regulatory, and residuals issues associated with water reuse applications are considered.

Chapter 12—Simultaneous Compliance: The use of ion exchange can impact the finished water by adding anions and cations and potentially affect the corrosivity and pH of the finished water. This chapter describes the potential impacts of ion exchange on distribution system water quality; presents an example of a distribution system water quality monitoring plan; and discusses experimental methods to evaluate the impact of ion exchange on distribution system water quality.

Chapter 13—Piloting: The use of pilot testing is important to prove the effectiveness of ion exchange for a specific application, to understand treatment challenges, and assess potential unintended consequences that could be caused by the use of ion exchange. This chapter discusses technical and logistical considerations of setting up and running a pilot testing program.

Chapter 14—Instrumentation, Commissioning, and Operations: A discussion of instrumentation, start-up, and operation and management (O&M) of ion exchange systems is provided.

Chapter 15—Case Studies: Case studies are provided for the removal of nitrate, PFAS, radionuclides, and DOC. The case studies describe projects where ion exchange has been implemented for the removal of these contaminants. The case studies include a description of the plant capacity, the raw water quality characteristics, and a description of the ion exchange system that was used.

CHAPTER 2

Fundamentals of Ion Exchange

The basic concepts that are described in this chapter include

- Overview of ion exchange
- Ion exchange resin properties and types
  - Ion exchange equations
  - Resin types: functional group and the mobile ion being exchanged
  - Applications of the basic four resin types (SAC, WAC, SBA, WBA)
  - Ion exchange kinetics
  - Selectivity and chromatographic peaking
  - Physical–chemical properties of resin
- Operating characteristics
  - Pretreatment requirements
  - Description of operating phases
  - Series and parallel operation
  - Water quality analysis
  - Sizing parameters
  - Equipment configuration guidance
  - Water quality monitoring related to operation
- Single-use resins
  - Applications for single-use resin

2.1 Overview

Ion exchange is a physical–chemical process where dissolved ions are removed from solution and exchanged for different ions of the same or similar electrical charge that are attached to the functional groups on the resin. In water purification, electrically charged particles in the aqueous solution (contaminants) are exchanged for ions on the resin that do not pose health risks. Many contaminants in source waters have either a positive or negative charge. Positively charged ions are referred to as cations (e.g., Na^+) and negatively charged contaminants are referred to as anions (e.g., Cl^-). Common contaminants include but are not limited to:

Cations

- Calcium and magnesium (hardness) Ca^{+2}, Mg^{+2}
- Iron and Manganese, Fe^{+2}, Mn^{+2}
- Barium, Ba^{+2}
- Strontium, Sr^{+2}
- Radium, Ra^{+2}

Anions

- Nitrate, NO_3^-
- Perchlorate, ClO_4^-
- Poly- and perfluoroalkyl substances (PFAS)*
- Total organic carbon or natural organic matter (TOC, NOM)
- Uranium, $UO_2(CO_3)_2^{-2}$, $UO_2(CO_3)_3^{-4}$
- Arsenic (arsenate, AsO_4^{-3})
- Chromate (CrO_4^{-2})

* The US Environmental Protection Agency (USEPA) has set a Health Advisory (HA) level for PFOS and PFOA compounds. Additional information on PFAS treatment can be found in Chapter 10—PFAS Removal.

This ion exchange process takes place within a vessel, or column, filled with specialized resin. The resins are spherical beads composed of functional copolymers and designed to be highly selective for the contaminant. The ion on the resin functional group defines the ionic form of the resin. A softening resin, for example, has positive sodium ions; therefore, it is referred to as being in the sodium ionic form. Depending on the "ionic form," the ion is exchanged for the ionic contaminant in the water phase. The contaminant in the water phase is retained on the solid phase resin. Figure 2-1 depicts how this process works. Ion exchange technology has many applications, but when used for drinking water purification, it is important to select a resin that is certified to NSF/ANSI-61 standard.

2.1.1 Mass Transfer

An ion exchange vessel will operate in three zones: exhausted zone, mass transfer zone, and fresh resin zone. A clean vessel starts with all fresh resin; raw water enters from the top, exhausting the resin from the top down until the entire vessel becomes exhausted. The ion exchange process takes place in the mass transfer zone. Mass transfer consists of three steps:

1. Diffusion of the ions from the bulk phase through the liquid
2. Diffusion of ions into the micropores of the ion exchanger
3. Ion exchange reaction

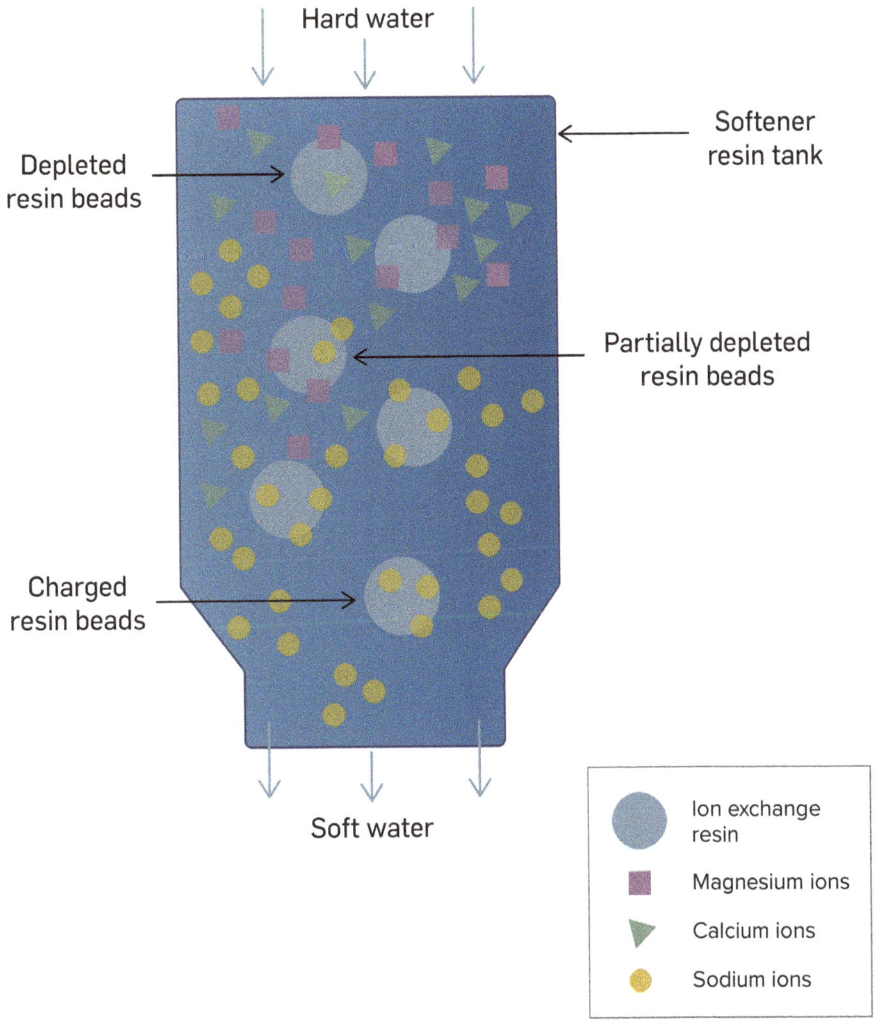

Figure 2-1 Ion exchange process for water softening

The exchange of the ion in the water phase onto the resin occurs in the mass transfer zone. The mass transfer process is depicted in Figure 2-2. The mass transfer zone decreases gradually until the bed is completely exhausted and breakthrough occurs. Breakthrough is the appearance of the unwanted contaminant at an unacceptable concentration in the effluent.

Figure 2-3 depicts the concentrations of the removed ions in the effluent verses the effluent volume. This plot is referred to as a breakthrough curve.

2.1.2 Resin Capacity and Regeneration

The total capacity of a resin represents the number of exchange sites available. Useful capacity, or what is referred to as operating capacity, is the fraction of the capacity that is economically accessed based on the raw water chemistry and regenerant dosage. The operating capacity is expressed as equivalents per liter (eq/L) or kilograins per cubic foot (kgr/ft^3) as $CaCO_3$ (1 eq/L = 21.85 kgr/ft^3). When a

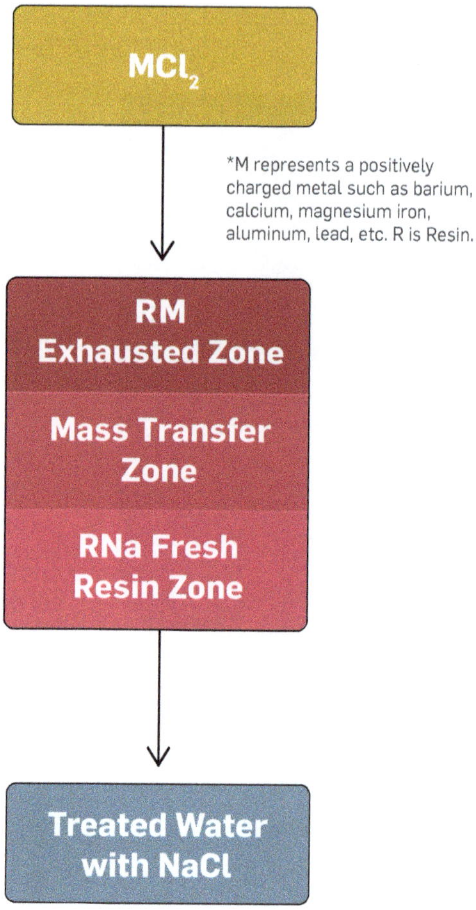

Figure 2-2 Diagram of ion exchange mass transfer

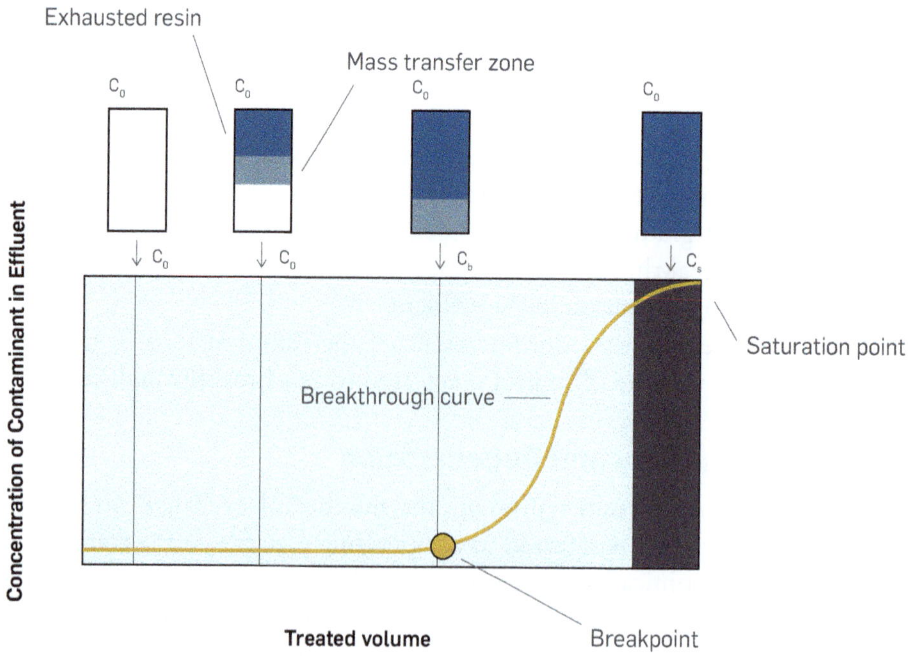

Figure 2-3 Example breakthrough curve

resin has reached its useful capacity, it needs to either be replaced (single use) or regenerated. Not all exchange sites are used during an exhaustion cycle nor are they all regenerated back to the initial ionic form during regeneration.

Regeneration is the process of removing the ions that have been picked up during service cycle and replacing them with a salt or other regenerant chemicals to restore the resin capacity. The flow through the media bed is reversed and a regenerant solution, typically a brine, is pushed through the bed. Additional information on the regeneration process and operations can be found in Section 2.3.

Regenerant dosage and concentration will have a direct impact on operating capacity and leakage. High dosage will increase operating capacity and reduce leakage. Breakthrough curves are available from resin manufacturers to define the relationship.

2.2 Ion Exchange Resin Properties and Types

A resin is a synthetic media, typically spherical, and generally 40–60 percent water by weight. The media is often referred to as "beads" and range in size from about 300–1,200 μm, or 50–16 US mesh. The resin consists of a polymer backbone, stationary functional group, and the transferable ion.

The two-step process used in ion exchange resin manufacturing includes polymerization of the copolymer followed by activation, where a functional group is attached to the copolymer. Functional groups such as sulfonic acid, amine and carboxyl provide the ion exchange property where ions in water are exchanged for counterions on the functional group. Selectivity of the ion exchange resin is governed by both molecular weight and valence.

The chemical composition of most resin's backbone is either a copolymer of polystyrene and di-vinyl benzene (DVB) crosslinked (strong acid cation [SAC], strong base anion [SBA], weak base anion [WBA]) or polyacrylic crosslinked with DVB.

Figure 2-4 depicts the chemical structure of styrene, DVB, and crosslinked polystyrene.

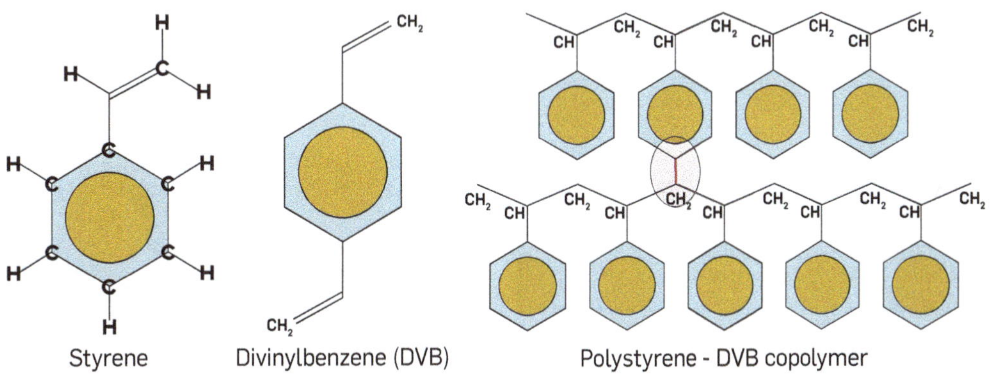

Figure 2-4 Chemical structure of resin

Resin can be either gel or macroporous. Gel resins are translucent while macroporous resins are opaque. Macroporous resins are made by adding a "porogen" during polymerization. The polymers form around the porogen forming the macropores, and the porogen is rinsed out after polymerization is complete. It should be noted that the matrix around the macropores is gel phase with small pores, so the macropores act as transport facilitators to allow the water to reach the active sites deeper in the bead where the ion exchange takes place. Figure 2-5 provides a close-up photograph of each of these types of resins.

While gels have very small pores, 10–20 Å (angstrom), macroporous resins have larger pores, 100–1000 Å. The angstrom is a unit of length equal to 10^{-10} meter, or 0.1 nanometer.

Figure 2-6 shows the relative size of gel resin beads compared to a pinhead.

Figure 2-5 Photograph of gel resin (left) and macroporous resin (right)

Figure 2-6 Relative size of gel ion exchange beads (purolite resin)

In most instances, the ion exchange process uses a synthetic resin that can be regenerated and used again. However, there are cases where economics of the contaminant dictate that the ion exchange resin is used once and disposed of. These resins are referred to as "single use" resins and are described later in this chapter.

Synthetic ion exchange resins and their function may be described as follows:

1. The resin is a spherical bead made of a polymer matrix (or backbone) to provide strength and structure. The beads are sized to minimize pressure drop (300–1,200 microns).
2. Stationary functional groups affixed to the copolymer matrix are the ion exchange sites. The functional group can be made up of positively charged (cationic) or negatively charged (anionic) ions. This determines the classification of the resin:
   a. SAC, typically in the Na^+ ionic form
   b. Weak acid cation (WAC), typically in the Na^+ or H^+ ionic form
   c. SBA, typically in the Cl^- ionic form
   d. WBA, free base or protonated form

   The acid and base in the resin description above refers to the strength of the functional group.
3. The mobile exchange ion is the ion that comes off the resin in exchange for the contaminant ion.

2.2.1 Ion Exchange Equation and Example

Removal of hardness or water softening is a common use of ion exchange technology. Using the softening resin example below, water is passed through the softening resin in the sodium form and calcium and magnesium are captured in the resin and exchanged for the sodium ions.

Equation 2-1, Service Mode, represents calcium removal by a SAC in sodium form:

$$2R\text{-}Na^+ + CaCl_2 \rightarrow R_2\text{-}Ca^{+2} + 2NaCl \qquad \text{(Eq 2-1)}$$

where R is the ion exchange resin and Na^+ is exchangeable ion on the functional group

As shown in Equation 2-1, sodium ions are exchanged for the calcium ions. This equation format will be used throughout this book and enables an understanding of how the effluent water quality will change as a result of the ion exchange process. Because the equation is reversible, the resin can be regenerated as in Equation 2-2, Regeneration Mode. In this example, a SAC resin is being regenerated with a 10 percent by weight concentration of sodium chloride (NaCl).

$$R_2\text{-}Ca^{+2} + 2NaCl \rightarrow 2R\text{-}Na^+ + CaCl_2 \qquad \text{(Eq 2-2)}$$

The ion exchange process exchanges a chemically equivalent amount of ions to the ionic contamination removed. However, the ion exchange process can have the unintended consequence of increasing water corrosivity. For example, alkalinity removed by a SBA resin in chloride form will depress pH to an acidic corrosive condition. At the same time, the SBA will also exchange the chloride ion for alkalinity, sulfate, and nitrate and increase the ratio of chloride to sulfate in the finished water. The chloride-to-sulfate molar ratio, also known as the CSMR, is a strong indicator of water corrosivity.

Ion exchange operates until the concentration of the contaminant of concern in the effluent water reaches its target, which is typically referred to as "breakthrough." When breakthrough occurs, there are two options: (1) regeneration or (2) to remove and replace the resin with new resin. Section 2.4 provides information on single use resins. The regeneration option is typically more economical when raw water concentrations are at the mg/L level. NaCl is a common regenerant, and the spent regenerant will require disposal. In many instances, a sodium chloride brine regenerant can be reused up to five times before disposal. Each of the contaminant chapters that follow will discuss waste-handling options associated with regeneration or spent-resin disposal.

2.2.2 Resin Types—Functional Groups and Reactions

The functional group attaches to the crosslinked copolymer backbone and determines the type of ion exchange resin. The four main types of resin and their functional groups are SAC, WAC, WBA, and SBA.

2.2.2.1 Strong Acid Cation

SAC exchange resins can generally remove all regulated cations regardless of if they are in the carbonate or noncarbonate form.

Equation 3 represents SAC-Na for cation removal by exchanging Na.

$$2R\text{-}SO_3\ Na^+ + M^{+2}Cl_2^- \rightarrow R_2\text{-}SO_3\ M^{+2} + NaCl \qquad \text{(Eq 3)}$$

M* is a positively charged metal such as barium, calcium, magnesium, iron, aluminum, lead, etc.

2.2.2.2 Weak Acid Cation

WAC resins are suitable for removal of regulated cations when operating in the hydrogen form as long as the cations are in the bicarbonate form such as temporary hardness, $Ca(HCO_3)_2$.

Regeneration is a single-step process where the WAC is regenerated with acid.

WAC resin can operate in either the H^+ or Na^+ form.

Equation 4a represents WAC-H for calcium removal by exchanging H.

$$2R\text{-}COOH + Ca^{+2}\left(HCO_3^-\right)_2 \rightarrow \left(R\text{-}COO^-\right)_2 Ca^{+2} + 2[H_2CO_3]_2 \qquad \text{(Eq 4a)}$$

Equation 4b represents WAC-Na for heavy metals by exchanging Na using Ca as an example.

$$2R\text{-}COO^{-}Na^{+} + Ca^{+2}Cl^{-}_{2} \rightarrow \left(R\text{-}COO^{-}\right)_2 Ca^{+2} + 2\left[Na^{+}Cl^{-}\right] \qquad \text{(Eq 4b)}$$

2.2.2.3 Strong Base Anion

SBA resins are suitable for the removal of all regulated anions including some fraction of TOC as well as for the removal of many PFAS compounds.

Resins contain an amine group—either trimethyl ammonium, which is called a Type 1 SBA, or dimethylethanolamine, which is called a Type 2 SBA. Both operate in the chloride form for potable water applications.

- Type 1, trimethyl ammonium, can impart an odor to the treated water if used intermittently.
- Type 2, dimethylethanolamine, is a better option for intermittent use due to low odor concerns versus Type 1 resins.

Equation 5a represents SBA for nitrate removal by exchanging OH.

$$R_4N^{+}OH^{-} + NaNO_3 \rightarrow R_4N^{+}NO_3^{-} + NaOH \qquad \text{(Eq 5a)}$$

Equation 5b represents SBA for nitrate removal by exchanging chloride.

$$R_4N^{+}Cl^{-} + NaNO_3 \rightarrow R_4N^{+}NO_3^{-} + NaCl \qquad \text{(Eq 5b)}$$

2.2.2.4 Weak Base Anion

WBA in the protonated form is suitable for removal of oxyanions such as hexavalent chromium. The raw water pH in this case must be maintained between 5 and 5.5 for efficient removal.

WBA, also known as acid adsorbers, do not have exchangeable ions until the free base sites on the resin are protonated by an acid. After protonation, the chloride or sulfate ion is exchanged for the target contaminant as in the case of chromate.

Equation 6a shows the protonation step.

$$R\text{-}CH_2\text{-}N(CH_3)_2 + H^{+}Cl^{-} \rightarrow R\text{-}CH_2\text{-}NH^{+}(CH_3)_2^{+}\,Cl^{-}\,(\text{protonation}) \qquad \text{(Eq 6a)}$$

Equation 6b shows the subsequent ion exchange step.

$$2R\text{-}CH_2\text{-}NH{+}(CH_3)_2{+}Cl^{-} + Na\,CrO_4^{-2} \rightarrow 2R\text{-}CH_2\text{-}NH{+}(CH_3)_2{+}CrO_4 + Na^{+}Cl^{-} \qquad \text{(Eq 6b)}$$

2.2.3 Ion Exchange Kinetics

Ion exchange kinetics, or the rate at which the ion exchange process occurs, is an important consideration when designing an ion exchange system and selecting a resin. The kinetics, or reaction time, will have a significant impact on the size of the treatment process. The volume of resin will need to be increased if the kinetics of a given resin are slow relative to other options. Slower kinetics result in more resin and larger vessels; however, this needs to be compared to the unit cost of the resin. For example, one resin may require half the reaction time but may be four times more expensive on a unit–cost basis.

Important factors that impact kinetics and that the design engineer and resin supplier should quantify include

- contaminant concentration,
- temperature,
- resin particle size, and
- degree of cross linking and hydration of the resin.

Generally SAC exchange resins have faster kinetics than WAC and WBA resins for the contaminants regulated in drinking water.

2.2.4 Selectivity and Chromatographic Peaking

Selectivity refers to the order in which specific ions are preferred by a given resin. The selectivity of an ion is governed by valance or oxidation state and molecular weight. All raw water sources contain multiple ions, both cations and anions. A cation exchange resin will remove all cations to some degree and anion exchange resins will remove all anions to some degree. Resins are manufactured to preferentially remove the specific ion of concern; for example, radium selective resins will have a stronger ionic bond for radium than for other divalent cations. The affinity factor, AF, is the numerical value associated with selectivity. SAC resins are normalized to sodium, and SBA resins are normalized for chloride. For example, lead has an AF of 5.0, which means that at equal concentrations lead is preferred by the resin by a factor of 5 over sodium.

Selectivity and relative affinity are approximately the same for monovalent ions. For divalent ions, such as sulfate or calcium, the relative affinities and selectivity are not the same and will vary based on the total dissolved solids (TDS) of the water. Table 2-1 presents the affinity factors for typical SAC and SBA resins (Edzwald 2011).

A less preferred ion will be concentrated on the resin and can exit the column at a concentration higher than their influent concentration. This effect is known as chromatographic peaking or sloughing. Chromatographic peaking can be minimized by selecting the proper resin and can be managed by proper regeneration timing. Resins are developed so that the ion of most concern is preferentially removed to minimize the potential health and regulatory impacts of chromatographic peaking. An example is an SBA nitrate selective resin with triethyl amine functionality, which has higher selectivity for nitrate compared to sulfate. The selectivity is reversed for standard SBA Type I resins with trimethyl amine functionality where nitrate can be released at a higher concentration than the influent

Table 2-1 Typical SAC and SBA selectivity

Strong Acid Cation (SAC)		Strong Base Anion (SBA)	
Ra^{2+}	13.0	CrO_4^{2-}	100.0
Ba^{2+}	5.8	SeO_4^{2-}	17.0
Pb^{2+}	5.0	SO_4^{2-}	9.1
Sr^{2+}	4.8	$HAsO_4^{2-}$	3.5
Cu^{2+}	2.6	HSO_4^{-}	4.1
Ca^{2+}	1.9	NO_3^{-}	3.2
Zn^{2+}	1.8	Br^{-}	2.3
Fe^{2+}	1.7	SeO_3^{2-}	1.3
Mg^{2+}	1.67	HSO_3^{2-}	1.2
K^{+}	1.67	NO_2^{-}	1.1
Mn^{2+}	1.6	Cl^{-}	1.0
NH^{4+}	1.3	HCO_3^{-}	0.27
Na^{+}	1.0	OH^{-}	0.10
H^{+}	0.67	F^{-}	0.07

if overrun. To quantify the potential impacts of the chromatographic peaking, a comprehensive water quality analysis measuring all the anions and cations should be performed and provided to the resin supplier. The analysis should capture any changes to water quality that occur over time. Recommended water quality analyses are provided later in this chapter.

Guidelines regarding selectivity include the following:

1. The most preferred ion in the selectivity chart will have long runs and a sharp breakthrough curve. The opposite is true for the remaining ions in the water.
2. The effluent concentration of the pre-saturated ion on the resin decreases in the effluent as the total ionic concentration in the effluent must remain constant.

2.2.5 Physical–Chemical Properties of Resin

Understanding the properties of the resin is important for sizing the system as well as comparing resins.

SAC and SBA resins share some similar characteristics with respect to particle size and moisture content but are much different with respect to chlorine tolerance and backwash rate. SAC resins can generally tolerate a free chlorine residual of up to 0.20 mg/L without resin damage, whereas for SBA resins any chlorine residual can cause damage to the resin. Table 2-2 presents the properties for Purolite's SAC and SBA resins. The properties of the SAC and SBA resins can vary based on supplier and, in general, SAC and SBA resins have properties similar to the properties listed below.

Table 2-2 Typical SAC and SBA (Cl form) physical and chemical characteristics and operating parameters

Property	SAC Resin	SBA Resin
Particle size, micron	300–200	300–1200
Specific gravity	1.25	1.05
Moisture content, %	44–60	44–60
pH range	0–14	0–14
Maximum operating temp., °F	250	180
Minimum operating temp., °F	34	34
Maximum feed turbidity, NTU	<1	<1
Maximum feed total iron, mg/L	<1	<1
Maximum feed total manganese, mg/L	0.25	0.25
Maximum free chlorine, mg/L	0.20	0.0
Backwash rate, gpm/ft^3	4–5	1–3
Backwash period, min	10–15	10–15
Bed expansion (%)	50	50
Volume capacity, eq/L	2.0	1.3
Linear velocity, gpm/ft^2	6–20	6–20
Surface loading rate, gpm/ft^3	1–5	1–5

2.3 Operating Characteristics

Pretreatment prior to ion exchange is often needed to limit mineral scaling, surface clogging, and resin fouling. Pretreatment may include filtration or addition of chemicals to reduce these issues.

2.3.1 Pretreatment

Ion exchange treatment processes will achieve their best performance when designed to minimize sources of fouling of both the resin and the equipment. Pretreatment will likely be required in order to address one or more feedwater characteristics that may impact performance. The location of the ion exchange process within a treatment train where multiple unit processes are used can also help mitigate impacts on resin performance due to feedwater characteristics. The following are typical water quality issues that may need to be mitigated through pretreatment or unit process placement.

Total suspended solids. Solids buildup on the resin bed can lead to uneven flow distribution and loading within the bed due to blockage and channeling. Backwashing of resin media as a means to remove solids buildup during service is not recommended since it disturbs the mass transfer zone and may lead to early contaminant breakthrough. Disturbances of the resin bed once placed into service should be avoided. Therefore, pretreatment to provide suspended solids removal is recommended. Typically, suspended solids removal is achieved by providing prefiltration using bag filters or cartridge filters. Filter ratings of 5 or 10 microns are typically used, but a particle size distribution analysis to characterize the suspended solids for a given feedwater can help confirm the prefiltration requirements. Solids removal is particularly important for single use resin applications since they should not be backwashed.

Iron and manganese. Iron and manganese are common co-contaminants, particularly in groundwater sources. While low levels of iron and manganese that remain in solution can pass through the resin system without a significant impact, pretreatment should be considered, particularly if it is determined that particulate iron or manganese are present. Pretreatment should also be considered if dissolved iron is present at concentrations of 1 mg/L or greater, dissolved manganese is present at concentrations of 0.2 mg/L or greater, and the potential exists for oxidation of these metals to their insoluble forms before contact with the resin.

Oxidants. The presence of oxidants in the feedwater to resin systems can lead to breakdown of the resin media and can lead to nitrosamines in the media effluent. Anion resin treatment systems should be designed so that the feedwater is oxidant free. Oxidant removal options include using GAC or dechlorination chemicals such as sodium bisulfite, calcium thiosulfate, etc. Alternately, implementation of ion exchange treatment upstream of oxidant addition or relocation of oxidant addition downstream of ion exchange, where possible, should be considered.

Scaling. Resin system feedwater chemistries that have a tendency to scale can impact both equipment and the performance of the resin; this applies particularly to water chemistries with a Langelier Saturation Index approaching +1 or greater. When scaling potential is high, consideration should be given to alternate locations in the treatment train where scaling potential may be lower. For example, air stripping used for carbon dioxide and/or volatile organic compound (VOC) removal may be a unit process in a treatment train that can increase a water's scaling potential due to release of carbon dioxide and the subsequent shift in carbonate chemistry equilibrium. Implementation of ion exchange upstream of air stripping may be beneficial to the resin's performance in such a scenario. Dissolved carbon dioxide will not impact the performance of the resin, as long as it is kept in solution under pressure.

In general, the use of polar sequestrants prior to resin treatment is not recommended as these chemicals can reduce the capacity of the resin. Acid feed ahead of the resin to reduce scaling potential can be considered where scaling is a significant concern; however, it will increase anionic loading and reduce the removal capacity of the resin.

Total organic carbon. With the exception of TOC removal resins described in Chapter 4, the presence of TOC can foul and reduce the capacity of the resin. Where TOC levels exceed 2 mg/L, pretreatment to reduce TOC should be considered to maximize the capacity of the resin. Pretreatment options include GAC and brine regenerable organic scavenger resins.

Microbial contaminants. Microbial contaminants that are naturally in the water source or introduced due to equipment contamination can lead to media fouling and excessive head loss across the resin bed. Proper disinfection of equipment and piping along with care during transfer of fresh resin to the disinfected equipment will minimize the risk of microbial contamination of the system. When microbial contamination is present in the source water, disinfection using UV or an oxidant may be required to prevent growth on the resin bed. As discussed previously, care must be taken when using oxidants for disinfection so as to avoid damaging the resins.

Oil and grease. Oil and grease, if found in the source water, will foul resin if not removed.

After pretreatment, or if pretreatment is determined to not be necessary, water is directed through the ion exchange vessel. The volume of resin in the vessel is considered 1 bed volume or BV. Raw water is run through the ion exchange vessel until the resin is exhausted, at which point the vessel is taken offline and regenerated with a regenerant (for cation exchange resins, the regenerate is typically sodium chloride, NaCl) through a backwash procedure. Figure 2-7 shows a schematic of the ion exchange process.

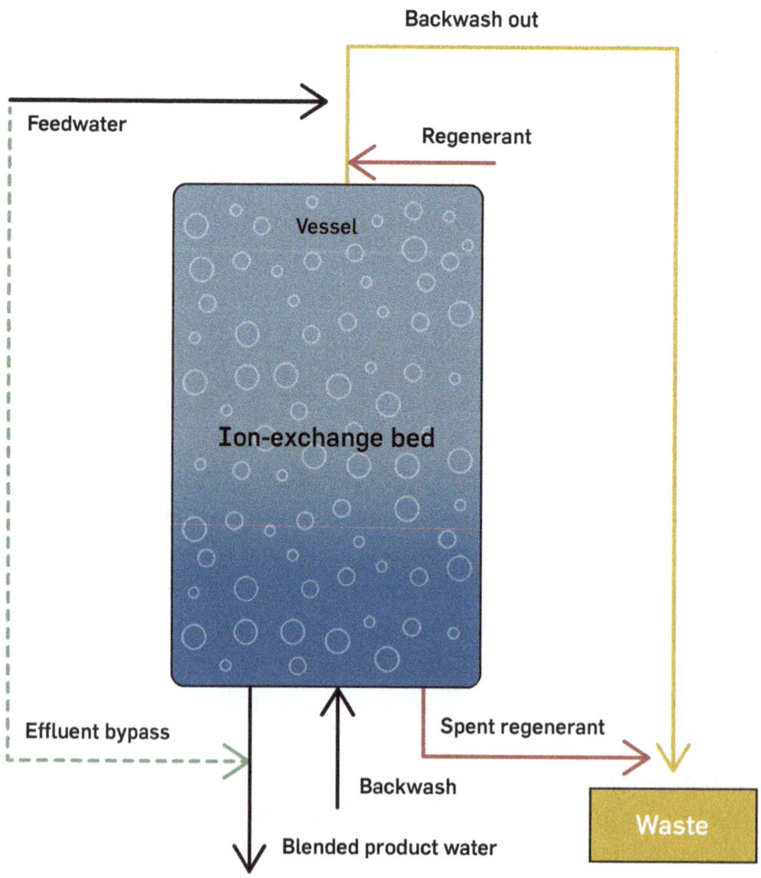

Figure 2-7 Ion exchange process schematic

2.3.2 Description of Operating Phases

The following describes the operating phases of a standard regenerable ion exchange system:

1. **Exhaustion (service cycle)**: in this mode, raw water is flowing through the resin, the contaminant is removed and exchanged for ion on the resin.
2. **Regeneration**
   a. Backwash (to waste): this is the first step in the regeneration cycle and the purpose of this step is to fluidize the bed and remove any particulates that pass through the pretreatment system. During backwash, the bed is typically expanded at least 50 percent.
      i. Backwash for co-current, also known as co-flow, is typically conducted every regeneration.
      ii. Backwash for counter current systems is conducted only when the pressure drop exceeds operating specifications—typically around 15 psi across a resin bed.

   b. Co-flow regeneration: the regenerant solution follows the same path as the solution to be treated. The flow rate ranges between 0.25 and 0.5 gpm/ft^3 of media. See Figure 2-8 and Table 2-3.
   c. Counter-flow regeneration: injection of the regenerant solution in the opposite direction of the service flow. The regenerant solution contacts the less exhausted resin layers first, making the regeneration process more efficient than co-flow regeneration. The flow rate ranges between 0.25 and 0.5 gpm/ft^3. See Figure 2-9 and Table 2-4.
   d. Split flow counter flow regeneration: the regenerant solution is injected through the top and bottom of the bed and exits through the intermediate collector, which is one third of the distance from the top of the bed. This is the most efficient regeneration. See Figure 2-10 and Table 2-5.

After the chemical addition, there will be excess regenerant on the bed that needs be removed to waste using a slow and fast rinse.

   e. Slow rinse (to waste): this is generally performed for 2–40 min (2–4 BVs).
   f. Fast rinse (to waste): this is generally performed for 10–20 min at the service flow rate (3–6 BVs).

NOTE: The hardness of the rinse water should be less than 10 gr/gal (171 ppm as $CaCO_3$).

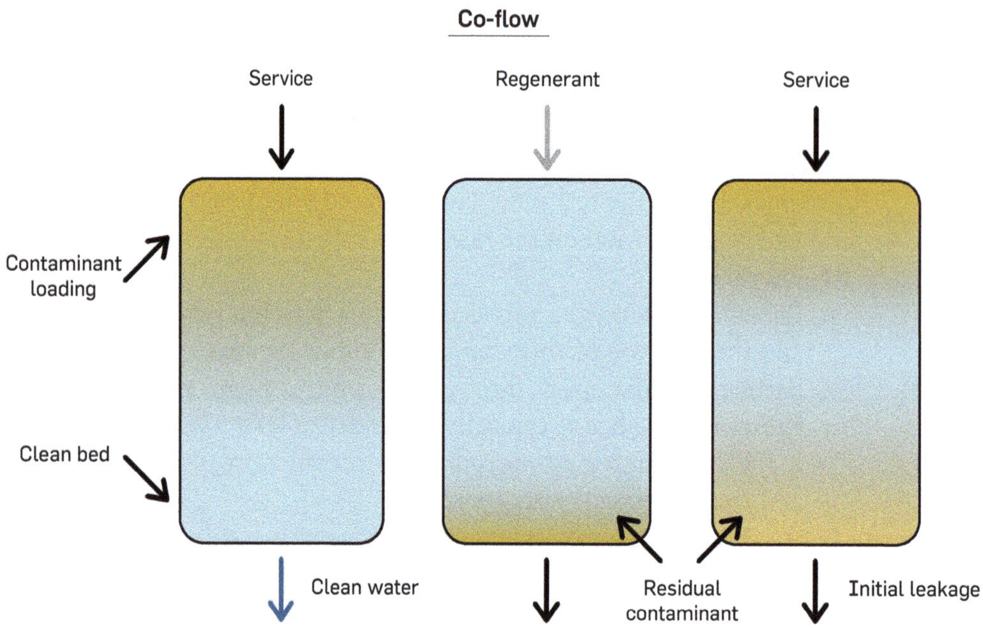

Figure 2-8 Co-flow regeneration

Table 2-3 Typical operating conditions for co-flow regeneration

Step	Design Basis	Duration
Backwash	Set for minimum water temperature to give 50% bed expansion	1 BV on clean water supplies and 2–3 BV where solids are present
Bed settle	To allow the bed to reform fully classified	3–8 min
NaCl injection	8–12 lb/ft^3 applied as a 10% solution at 2–4 BV/h (0.25–0.5 gpm/ft^3)	Typically 20–45 min depending on regeneration level and flow rate
Slow rinse	2–4 BV (0.25–0.5 gal/ft^3) at approx. regenerant flow rate	Typically 30–40 min depending on regeneration level and flow rate
Final rinse	3–6 BV (22.5–45 gal/ft^3) preferably at service flow rate or alternatively > 8–16 BV/h (60–120 gpm/ft^3)	Typically 10–20 min

BV: bed Volume, BV/h: bed volume per hour

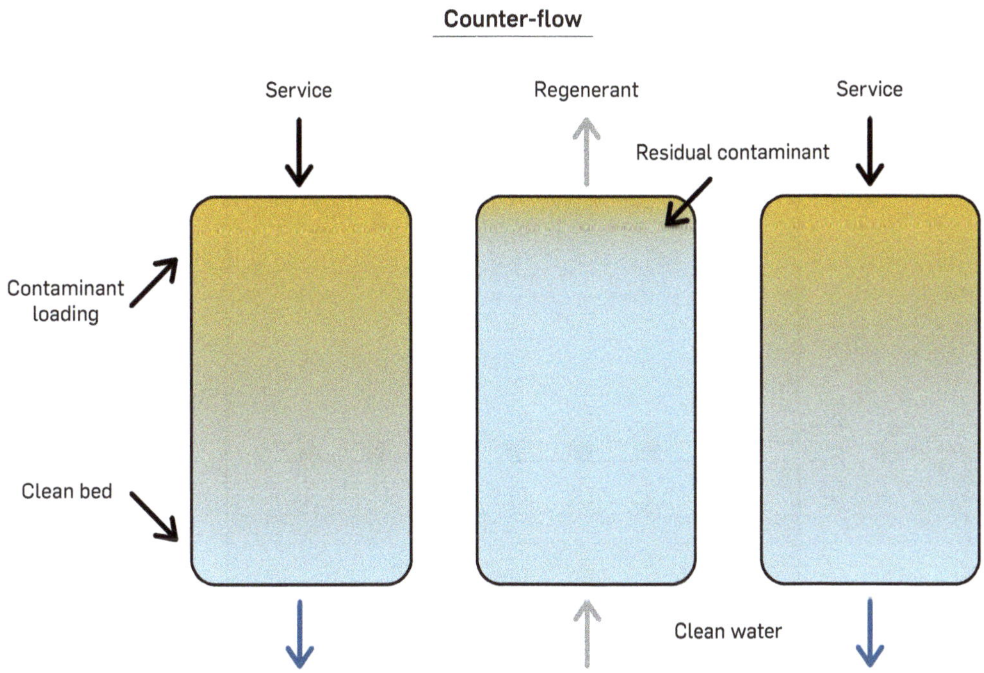

Figure 2-9 Counter flow regeneration

Table 2-4 Typical operating conditions for counter-flow regeneration

Step	Design Basis	Duration
NaCl injection	8–12 lb/ft^3 applied as a 10% solution at 2–4 BV/h (0.25–0.5 gpm/ft^3)	Typically 20–45 min depending on regeneration level and flow rate
Slow rinse	1–2 BV (7.5–15 gal/ft^3) at approx. regenerant flow rate	Typically 30–40 min depending on volume of water applied and flow rate
Final rinse	2–4 BV (15–30 gal/ft^3) preferably at service flow rate or alternatively 8–16 BV/h (60–120 gpm/ft^3)	Typically 10–20 min

BV: bed Volume, BV/h: bed volume per hour

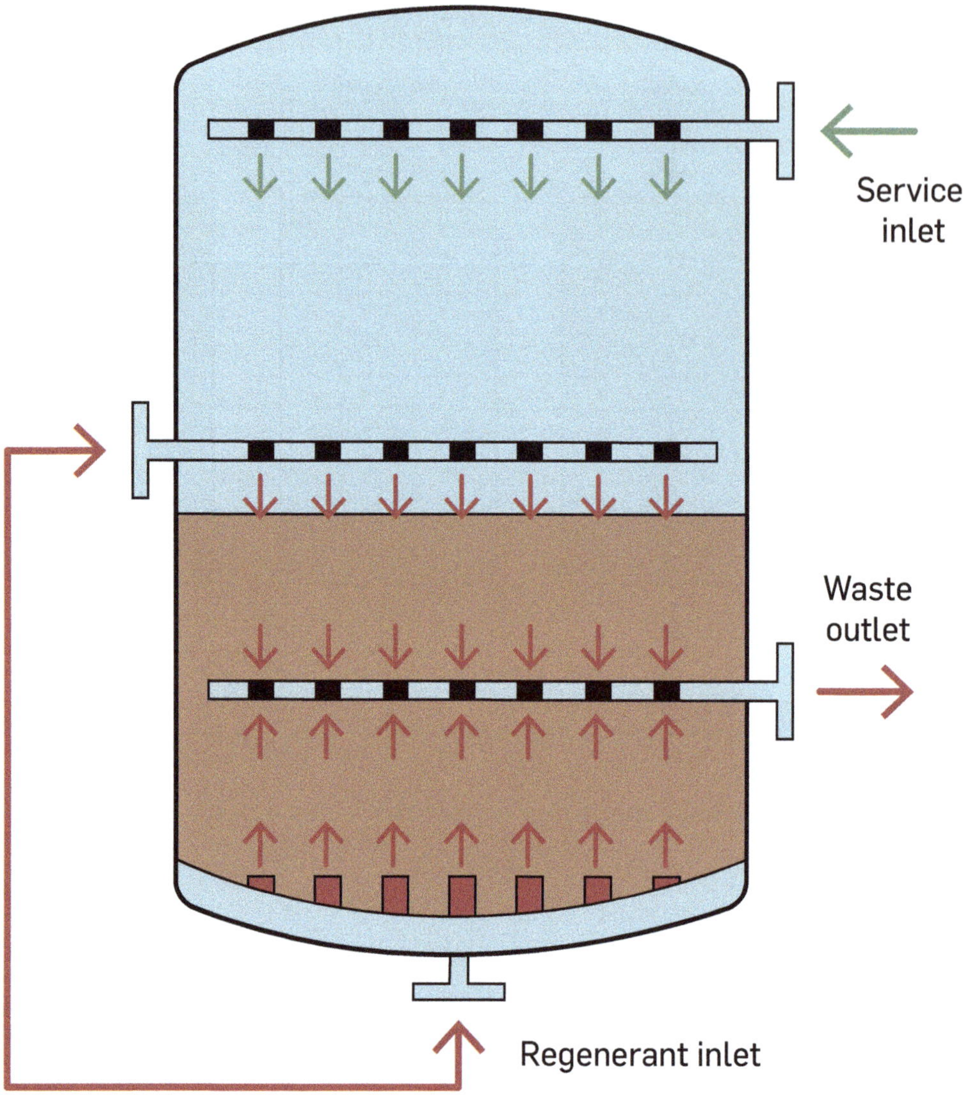

Figure 2-10 Split flow counter flow regeneration

Table 2-5 Split flow counter flow regeneration (split flow 1/3 down, 2/3 up)

	Step	Design Basis	Duration
1a and 1b occur simultaneously			
1a	NaCl injecion—Down flow (1/3)	48 g/L (3 lb/ft^3) applied as a 10% solution at 2–4 BV/h (0.25–0.5 gpm/ft^3)	Typically 20–45 min depending on regeneration level and flow rate
1b	NaCl injecion—Up flow (2/3)	96 g/L (6 lb/ft^3) applied as a 10% solution at 2–4 BV/h (0.25–0.5 gpm/ft^3)	Typically 20–45 min depending on regeneration level and flow rate
2a and 2b occur simultaneously			
2a	Slow rinse	1–2 BV (7.5–15 gal/ft^3) at approximate regenerant flow rate	Typically 20–45 min depending on volume of water applied and flow rate
2b	Slow rinse	1–2 BV (7.5–15 gal/ft^3) at approximate regenerant flow rate	Typically 20–45 min depending on volume of water applied and flow rate
3	Final rinse Down on entire vessel	2–4 BV (15–30 gal/ft^3) preferably at service flow rate or alternatively >15 BV/h (2 gpm/ft^3)	Typically 10–20 min

BV: bed Volume, BV/h: bed volume per hour

2.3.3 Series and Parallel Operation in SAC or SBA Regenerable Systems

Ion exchange vessels can be operated in series or parallel. In series operation, there are vessels both in service and standby, typically two in series flow (lead lag) and one in standby. The first vessel provides bulk removal of the contaminant, and the second vessel is used as a polisher to achieve the desired water quality. Determining the lead vessel change out is calculated using predictive modeling and or by taking grab samples for lab analysis. Operating as a "merry-go-round," the exhausted lead vessel will be replaced by the lag vessel, which still has adequate capacity, and the lag vessel is replaced with a fresh resin vessel.

Parallel operation is designed to operate with vessels in different stages of exhaustion. The effluent from all columns is blended, and standby fully regenerated resin vessels are placed in service as operating vessels reach exhaustion.

2.3.4 Water Quality Analysis—Anion and Cation Balance

An analysis of the feedwater is needed to determine the appropriate resin, as well as the pretreatment requirements. For feedwaters that vary, multiple sampling events should be conducted. When performing the analyses, the equivalency of the cations and anions must be equal. This is achieved by converting each "ion" to its calcium carbonate equivalent. Table 2-6 presents typical feedwater analyses that should be performed along with the factors needed to compute its calcium carbonate equivalence.

Table 2-6 Raw water analysis necessary for drinking water

Substance	**to $CaCO_3$ Multiplier**	**as Ion**	**as $CaCO_3$**
Calcium, Ca^{2+}	2.5		
Magnesium, Mg^{2+}	4.12		
Sodium, Na^{+}	2.18		
Potassium, K^{+}	1.28		
Barium, Ba^{2+}	0.73		
Strontium, Sr^{2+}	1.14		
Total Cations	—		
Bicarbonate, HCO^{3-}	0.82		
Carbonate, CO_3^{2-}	1.67		
Chloride, Cl^{-}	1.41		
Sulfate, SO_4^{2-}	1.04		
Nitrate, NO^{3-}	0.81		
Phosphate, PO_4^{3-}	1.58		
Fluoride, F^{-}	2.66		
Total Anions	—		
Additional Parameters to Assist in Ion Exchange Selection			
Total Hardness, mg/L as $CaCO_3$	—		—
Iron, Total	—		—
Iron, Dissolved	—		—
Silica, Total	—		—
Silica, Dissolved	—		—
Manganese, Total	—		—
Manganese, Dissolved	—		—
TDS	—		—
TSS	—		—
Hydrogen Sulfide	—		—
pH	—		—
TOC	—		—

2.3.5 Sizing Parameters

The parameters needed to size an ion exchange system consist of:

- Raw water quality—See Table 2-6
- Selection of the resin and its exchange capacity
- Empty bed contact time, typically a minimum of 1.5–2 min or 40 BV/h
- Linear velocity—typical design 8–12 gpm/ft^2
- Specific flowrate 1–5 gpm/ft^3
- Media depth—minimum 3 ft
- Configuration: lead/lag vessel (series) configuration
- Sizing regeneration system
  - Backwash supply and flow rate to achieve 50 percent minimum bed expansion
  - Brine volume, dosage, and flow rate
  - Displacement and fast rinse flow rate and volume
  - Spent waste volume
- A common form representing hardness concentration is grains per gallon (gr/gal)
  - 1 gr/gal = 17.1 ppm

An overview of these components follows, and the individual contaminant chapters contain design examples.

2.3.5.1 Empty Bed Contact Time

The empty bed contact time (EBCT) is the volume of the resin divided by the flow rate. The EBCT is resin and contaminant specific but generally ranges between 1.5 and 8 min. EBCT also represents a single bed volume, and bed volumes can be used to project the life span of the resin.

2.3.5.2 Linear Velocity/Surface Loading Rate

The linear velocity/surface loading rate (SLR) is the flow rate divided by the surface area of the vessel. Most ion exchange systems operate at an SLR of 6–12 gpm/ft^2. Higher SLRs result in smaller diameter vessels and higher pressure drop.

2.3.5.3 Media Depth

Media depth generally ranges between 36 and 72 in. The maximum depth is dependent on pressure drop allowances. The medial depth is used in conjunction with the vessel diameter to provide the volume of resin. Optimization is performed to provide the best combination of vessel diameter, pressure drop, and regeneration frequency.

2.3.5.4 Regeneration Chemicals and Concentrations

Regeneration chemicals can consist of sodium chloride, hydrochloric acid, sulfuric acid, and sodium hydroxide. The choice and selection of these chemicals depends on the type ion exchange system.

- For SAC using sodium as the exchange ion, a concentration 10 percent sodium chloride is generally used.
- For WAC using hydrogen as the exchange ion 4 percent HCl or 0.75 percent sulfuric acid is used. If operating in the sodium form, regeneration is first done with acid and then the resin is converted to the sodium form using caustic soda.
- For SBA using chloride as the exchange ion, a concentration of 8–12 percent sodium chloride is used.

Note that all these regeneration chemicals are dilute solutions, and softened water should be used for dilution.

Simple water softening example: Consider a well with 300 ppm hardness as calcium carbonate ($CaCO_3$). Knowing the flow rate will allow the volume of water that can be treated before regeneration is needed. The following example describes how the exchange capacity can be used in system design:

> A well has a design flow rate of 700 gpm. A resin with a capacity of 30 kgr/ft^3 and 200 ft^3 of resin is selected. The resin will have an operating capacity of 6,000,000 gr (30 kg/ft^3 × 1000 × 200 ft^3)
>
> The resin will also have an EBCT of 2.14 min EBCT (200 ft^3 × 7.48 gal/ft / 700 gpm).
>
> 300 ppm as $CaCO_3$/17.1 ppm as $CaCO_3$/gr =17.54 gr/gal
>
> 6,000,000 gr / (17.54 gr/gal) = 342,075 gal per cycle

As the mass transfer zone moves down through the column, eventually the bed will become exhausted, meaning all the exchange sites are occupied, and the contaminant will be detected in the effluent. The resin then will be regenerated or replaced.

2.3.6 Equipment Configuration Guidance

Best practices for ion exchange system equipment design include the following:

- Verify that all components be NSF 61–approved for drinking water applications.
- Vessels and treatment train configurations are sized to maintain recommended linear velocities over the anticipated range of system flows.
- Vessel internal free-board is adequate to accommodate resin backwashing, if required.
- Effluent collectors/underdrain should be designed to maintain uniform flow distribution.
- Design should include intermediate resin bed sample ports to allow monitoring of water quality within the bed. Typically, resin bed sample ports are located at 25, 50, and 75 percent of the bed depth. These samples as well as sample ports on the vessel inlet and outlet allow monitoring of contaminant loading and exhaustion rates and facilitate planning for resin changeouts.

- Resin traps in vessel effluent lines are sized to keep resin from escaping the ion exchange system should there be a failure in the vessel underdrain system.
- Design should include vessel inlet and outlet pressure gauges and differential pressure monitoring instrumentation.
- Design should include provisions for slurry loading of resin to prevent risk of biological contamination during resin installation.

2.3.7 Water Quality Monitoring Related to Operation of Regenerable Systems

2.3.7.1 General Functional Description

The operation of ion exchange process needs be terminated before the effluent exceeds the maximum contaminant level (MCL) or system specific water quality goal. Upon termination of the run cycle, the system is regenerated. The most typical method for determining when to regenerate the resin is to use the volume of water fed to the resin. The resin has a specific exchange capacity, and based on the raw water concentration of the ions, the volume of the water that can be treated can be computed as described previously. To allow for the appropriate factor of safety, as part of the commissioning processes, effluent samples should be obtained to calibrate the system.

2.3.7.2 Grab Samples

Influent and effluent sampling should be performed in accordance with the local regulatory requirements. In addition, a sampling plan should be developed. The specific parameters and frequency will depend on the application.

2.3.7.3 Online Instrumentation

Conductivity and pH are two online parameters that can be used as surrogates for monitoring performance. As discussed in Section 2.3.1, Pretreatment, high influent solids can quickly exhaust ion exchange resin, so monitoring influent water quality is important for resin maintenance. Conductivity is much easier to measure and can be related to TDS with the below relationship:

$$\text{TDS (mg/L)} = k \cdot \text{EC } (\mu\text{S/cm})$$

where EC is electrical conductivity, and k is the conversion factor, which is related to the chemical composition of the water. For typical natural waters such as stream and lake water, the value of the conversion factor is usually between 0.6 and 0.7, and a value of 0.64 is typical.

Chapter 14, Instrumentation, Commissioning, and Operations, contains additional information about online instrumentation and advancements regarding ion-specific analyzers.

2.4 Single-Use Resins

Analytical techniques have been developed to measure many contaminant concentrations at the part per billion (ppb) and part per trillion (ppt) levels. Adverse health effects for many contaminants are being discovered at these lower concentrations. Single-use resins are becoming more economically viable for many of these contaminants.

Single-use resins are designed to be extremely selective for trace contaminants in the water. These resins can last from months to years without replacement depending on background water chemistry. The operation of these single use systems is comparable to that of granular activated carbon where the system is operated to breakthrough and the media is replaced.

Single-use resins are disposed of when they are spent, so there is no regeneration or brine waste. Disposal options include landfill, waste-to-energy incineration, or encasing the spent resin in concrete. Some resin service providers will provide disposal service as part of a turnkey service. The appropriate disposal method will be based on issues, such as the contaminant to be removed, the characteristics of the spent resin, the availability of facilities that will handle the waste, local and national regulations, and cost. The contaminant chapters provide additional information regarding waste disposal.

Most of the contaminants that are amenable to single-use resins are anions, so SBA and WBA are the most common single-use resins. Contaminants that can be treated with single-use resins include chromate, uranium, perchlorate, radium, and perfluoroalkyl substances (PFAS). These are discussed in subsequent chapters.

As a resin is consumed, a mass transfer zone is established in the resin bed. The resin bed must not be backwashed after the initial loading when the vessel is put into service. Backwashing will disrupt the mass transfer zone, which can cause premature breakthrough. The resin is treating trace contaminants at part-per-billion or part-per-trillion levels, so even a small "bump" of the bed can cause early breakthrough.

2.4.1 Pretreatment

2.4.1.1 Total Suspended Solids

An important design consideration of a single-use resin system is proper prefiltration. Because these resins are in the field for extended periods of time, and because the bed cannot be backwashed, eliminating dirt and silt from the influent water is imperative. Properly sized prefilters should be installed and maintained on a regular basis. It is recommended to perform a particle size distribution analysis of the influent water and ensure the prefiltration is put in place.

If total suspended solids (TSS) do accumulate in the resin bed, it will cause channeling of the water, resin movement will occur, and the bed will experience early breakthrough.

2.4.1.2 Other Influent Considerations

There are several other considerations for single use resins. If any of the following are present in the water, pretreatment should be installed.

- Oil and grease—resin will be fouled by oil and grease and must be removed to non-detect.
- Iron and manganese—these metals can precipitate on downstream media.
- Oxidants—prolonged contact with oxidants such as chlorine, chloramine, hypochlorite, permanganate, or ozone will destroy the resin and increase the potential for nitrosamine formation in the effluent water.
- Scaling—Ensure the Langelier Saturation index (LSI) does not indicate scaling potential.
- Microorganisms—Microorganisms are often found in groundwater and in equipment in contact with the water. Careful investigation should be performed to identify the types, source, and concentrations.
- Total organic carbon—TOC is naturally present in most groundwater and surface water. Often, TOC has negatively charged components in the mix of organics. If the TOC is over 2 ppm, TOC can foul the resin and negatively impact performance. The extent of fouling depends on the molecular weight distribution of the TOC and its anionic characteristics.

CHAPTER 3
Hardness Removal

3.1 Introduction

Hardness is a common water quality parameter typically referring to the amount of dissolved calcium and magnesium in the water. Hardness includes all dissolved multivalent metallic cations in the water including iron, manganese, and aluminum; however, this chapter will focus on calcium and magnesium hardness.

Calcium and magnesium are naturally occurring and are usually found in groundwater sources. Hardness itself does not have adverse health effects at concentrations typically found in groundwater, but hardness will have adverse effects on boiler and heating and cooling system performance when precipitated as carbonate, phosphate, or sulfate scales on heat transfer surfaces. The performance of nanofiltration and reverse osmosis membranes used in water reuse or desalination systems will be affected in the same way. Hardness can also have impacts for the consumer. Hard water requires more soap to produce a lather, affecting laundry, shampooing, and dish washing.

This chapter describes how ion exchange systems can be used for the removal of naturally occurring hardness in groundwater, including:

- properties and chemistry associated with hardness in groundwater sources,
- drinking water regulations regarding hardness (US Environmental Protection Agency [USEPA] and the World Health Organization [WHO]),
- source water quality parameters that can impact the selection of ion exchange resins suitable for hardness removal,
- an overview of the types of softening resins that can be used for hardness removal,
- an overview of co-flow and counter-flow softener operation,

- design guidelines including calculations to determine capacity and residuals generation, and
- one design example.

3.2 Hardness Occurrence and Chemistry

The primary source of hardness in water is dissolved metallic ions from rocks, seepage, and runoff from soils. Calcium and magnesium are the most common metals in hardness, but as mentioned, the definition of hardness includes all multivalent metals such as iron, manganese, barium, strontium, aluminum, and zinc.

Hardness is the sum of the metallic cations that are present as neutral salts:

- Calcium, Ca^{+2}
- Magnesium, Mg^{+2}
- Iron, Fe^{+2}
- Manganese, Mn^{+2}
- Barium, Ba^{+2}
- Strontium, Sr^{+2}
- Aluminum, Al^{+3}

We will continue referring to hardness associated with calcium and magnesium. The divalent metals are bound to a negatively charged ion to form a neutral compound or salt. The metals bind to sulfate, nitrate, or chloride to form a salt. These compounds are known as permanent hardness. When metallic cations bind with bicarbonates, those compounds are known as "acid salts" or temporary hardness.

- **Salts (permanent hardness)**
  - Ca^{+2} or $Mg^{+2}\,SO_4^{-2}$
  - Ca^{+2} or $Mg^{+2}\,(Cl^-)_2$
  - Ca^{+2} or $Mg^{+2}\,(NO_3^-)_2$
- **Acid salts (temporary hardness)**
  - Ca^{+2} or $Mg^{+2}(HCO_3^-)_2$

Hardness ion concentration is typically expressed in either ppm as ion (ppm Ca^{+2} or Mg^{+2}) or ppm as calcium carbonate (ppm as $CaCO_3$). Laboratories typically report hardness and alkalinity in mg/L as $CaCO_3$, so it is important to understand how to convert from one reporting value to another, ppm as Ca^{+2} or ppm Ca^{+2} as $CaCO_3$.

- To convert from mg/L as metallic ion, to mg/L as $CaCO_3$, divide the equivalent weight (molecular weight divided by valence) of $CaCO_3$ by the equivalent weight of the ion, then multiply the result by the mg/L as ion.
- Example: convert 100 mg/L Ca^{+2} as ion, to calcium as mg/L as $CaCO_3$:
  - Calculate the molecular weight (MW) of calcium carbonate:

$$MW = 40(Ca) + 12(C) + 3\times16(O) = 100$$

- Calculate the equivalent weight (EW) of calcium carbonate by dividing MW by valence:
  EW of calcium carbonate = MW / valence = 100 / 2 = 50
- Calculate the EW of Ca^{+2}:
  MW of Ca = 40
  EW = MW / valence = 40 / 2 = 20
- Divide the EW of $CaCO_3$ by the EW of Ca^{+2}:
  50 / 20 = 2.5
- Multiply the ratio of EW of $CaCO_3$ to EW of Ca^{+2} by the concentration of Ca^{+2}:
  2.5 × 100 mg/L of Ca^{+2} as ion = 250 mg/L Ca^{+2} as $CaCO_3$.

The same mathematical approach can be applied to any ion by converting from one reporting value to another. When expressed as calcium carbonate, all ions are on an equivalent basis and may be added to determine total cations and total anions, which must balance. This is where the term "balanced water analysis" comes from. A balanced water analysis is recommended.

3.3 Hardness Regulations and Guidelines

As there are no known health effects associated with calcium and magnesium at the levels typically found in drinking water sources, regulations are limited to secondary standards (aesthetic non health effects) as well as guidance from AWWA and WHO. As described earlier, the major non health impact of hardness is associated with scaling on appliances and piping. Some metals in hardness (iron and magnesium) can also impart an acrid flavor to the water.

3.3.1 World Health Organization Guidelines

WHO does not have specific regulations but has established a guideline for hardness that is contained in WHO/HSE/WSH/10.01/Rev/1 titled "Background document for development of WHO Guidelines for Drinking-water Quality." This document provides the following classifications and concentrations regarding hardness (WHO 2011).

Hardness Level	Total Hardness (as mg/L $CaCO_3$)
Soft	<60
Moderately hard	60–120
Hard	120–180
Very hard	>180

The WHO guideline also provides additional information and criteria to assist the utility in determining whether to provide treatment, and this information includes taste impacts, scaling impacts, and corrosion impacts.

3.3.2 AWWA

According to AWWA's *Water Quality and Treatment* 6th edition, the following classifications are listed.

Hardness Level	Total Hardness/L (as mg/L $CaCO_3$)
Soft	<75
Moderately Hard	75–150
Hard	150–300
Very Hard	>300

(Edzwald 2011)

3.3.3 Other Agencies

The New Jersey Department of Environmental Protection (NJDEP) has established a secondary standard of 250 mg/L as $CaCO_3$ for hardness and recommends softening ion exchange to remove hardness below the standard. This may have limitations because the NJDEP established a secondary standard of 50 mg/L for sodium. More information regarding sodium cycle ion exchange is discussed later in this chapter.

The Minnesota Rural Water Association recommends that potable water "should have a total hardness of less than 75 to 85 mg/L as $CaCO_3$ and a magnesium hardness of less than 40 mg/L as $CaCO_3$ to minimize scaling at elevated temperatures."

Many water utilities choose to operate their systems at values ranging between 75 and 300 mg/L as $CaCO_3$, and the decision as to whether to remove hardness is generally based on the following factors:

- Severity of impact on water heaters
- Level of customer complaints
- Ability of the system to dispose of residuals generated from hardness removal

3.3.4 Additional Hardness Regulations

As described in Section 3.2, hardness is the sum of the metallic cations that are present as neutral salts. While this chapter focuses on calcium and magnesium, there are additional metals that can contribute to hardness. Regulators such as WHO and USEPA have guidance for some of these metals—summarized in Table 3-1. If these metals are found, resin selection should be evaluated to assess the impact of these metals.

Table 3-1 Hardness metals regulations/guidance

Metal	Agency	Type of Limit/Guidance	Limit/Guidance
Strontium	USEPA	Lifetime HA	4 mg/L
Barium	USEPA	MCL	2 mg/L
Manganese	USEPA	SDWR	0.05 mg/L
Aluminum	USEPA	SDWR	0.05 – 0.2 mg/L
Iron	USEPA	SDWR	0.3 mg/L

HA: health advisory, MCL: maximum contaminant level, SDWR: secondary drinking water regulation

3.4 Ion Exchange Resin Treatment Options for Hardness Removal

Hardness removal can be accomplished using three methods:

- Precipitation and clarification
  - through the addition of lime and soda ash
  - using a pellet softening system
- Nanofiltration
- Ion exchange

Before the development of nanofiltration membranes, precipitation and clarification have historically been used. The system involves chemical addition and the use of circular clarifiers. Settled sludge was either treated in a lagoon or mechanically dewatered. This technology is suitable for the range of flow rates encountered in municipal water systems.

Pellet softening involves the precipitation of hardness on a sand particle and forms a calcium carbonate pellet. Soda ash or lime is injected at the base of an upflow vertical reactor in which a bed of sand is maintained in fluidization. Hardness precipitates on the sand and forms a bead or pellet shape. This pellet can be used by concrete manufacturers and reduces residuals handling. This technology is generally suitable for smaller flow rates (5 mgd or less) depending on economics.

Nanofiltration uses a spiral wound membrane and operates in the same manner as reverse osmosis. Nanofiltration will generate a continuous brine stream that is disposed of in a sewer, by deep well injection, or to a saline water body.

Ion exchange for hardness removal uses a strong acid or weak acid cation exchange. Softening ion exchange systems used for hardness removal are also capable of removing iron and manganese.

Ion exchange systems are generally used for smaller municipal well systems that have a capacity of 5 mgd or less. Softeners are also used on private wells as point-of-entry (POE) systems for home and institutional drinking water.

The two types of cation exchange resins used for hardness removal:

- Strong acid cation (SAC) exchange, and
- Weak acid cation (WAC) exchange.

SAC resins can remove more than 95 percent of the calcium and magnesium typically present in drinking water. WAC resins remove only the carbonate hardness.

To select the most appropriate resin for hardness removal, the concentrations of cations and anions of the source water as a "balanced water analysis" must be identified, as well as the desired finished water hardness specification. The composition and operating characteristics of both SAC and WAC resins are described in the following section. Guidelines are provided for selecting between SAC and WAC resins.

3.4.1 Strong acid cation (SAC) Exchange Resins

As described in Chapter 2, Fundamentals of Ion Exchange, SAC resins have negatively charged sulfonic acid functional groups to which the exchangeable cations are attached. Both temporary (carbonate) and permanent (non-carbonate) hardness can be removed in a SAC ion exchange system. Equation 3-1 describes the SAC removal process.

- SAC in the sodium form showing calcium being exchanged:

$$2R\ SO_3\text{-}Na^{+} + CaCl_2 \rightarrow 2R\ SO_3\text{-}Ca^{+2} + 2NaCl \qquad \text{(Eq 3-1)}$$

SAC resins in the sodium form will add 2 mg/L of sodium (3.33 mg/L as $CaCO_3$) for every 1 mg/L of calcium (2.5 mg/L as $CaCO_3$) removed. As described previously, USEPA does not have a maximum contaminant level (MCL) for sodium, but the NJDEP has a secondary standard of 50 mg/L for sodium. Many groundwaters have some background sodium in the water, and if it is desired to stay below a sodium concentration of 50 mg/L, a maximum of 25 mg/L as $CaCO_3$ of hardness can be removed. If this cannot be achieved, the WAC operated in the hydrogen form should be considered (see next section). SAC resins in the sodium form that are used for hardness removal generally have the properties shown in Table 3-2.

Consideration should be given to the typical operating recommendation shown in Table 3-3.

Table 3-2 SAC resin typical physical and chemical characteristics

Characteristics	Description
Polymer structure	Polystyrene cross-linked with DVM
Physical form	Amber, clear spherical beads
Functional groups	Sulfonic
Ionic form, as shipped	Na^+
Total capacity, Na^+ form	2.0 eq/L (43.7 kgr/ft^3) min.
Moisture retention, Na^+ form	44–48%
Particle size range	300–200m; 1% max. <300m
Uniformity coefficient	1.7 max.
Reversible swelling	8% max.
Specific gravity, Na^+ form	Approx. 1.29
Shipping weight	800–840 g/L (50.0–52.5 lb/ft^3)
Maximum temperature limit	120°C (250°F)

DVM: di-vinyl benzene (see Chapter 2)

Table 3-3 SAC typical operating conditions for hardness removal

Step	Design Basis	Duration
Backwash	Set for minimum water temperature to give 50% bed expansion	1 BV on clean water supplies and 2–3 BV where solids are present
Bed settle	To allow the bed to reform fully classified	3–8 min
NaCl injection	50–250 g/L (3–15 lb/ft^3) applies as a 10% brine solution at 2–4 BV/h (0.25–0.5 gpm/ft^3)	Typically 20–45 min, depending on regeneration level and flow rate
Slow rinse	1–3 BV (7.5–22.5 gal/ft^3) at approx. regenerant flow rate	Typically 20–45 min, depending on regeneration level and flow rate
Final rinse	3–6 BV (22.5–45 gal/ft^3) preferably at service flow rate or alternatively >15 BV/h (2 gpm/ft^3)	Typically 10–20 min

BV: bed volume, BV/h: bed volume per hour

3.4.2 Weak Acid Cation Exchange Resins

WAC resins have a carboxylate functional group and remove only temporary hardness as shown in Eq. 3-2.

$$2\ R\text{-}COOH + Ca(HCO_3)_2 \rightarrow (RCOO)_2\,Ca + 2H_2O + 2CO_2 \quad \text{(Eq 3-2)}$$

(Pingale 2005)

Carbon dioxide gas is generated as part of the WAC process. The dissolved carbon dioxide will reduce the pH of the water and can be removed by either a forced gas or vacuum decarbonator.

WAC resins that are used for hardness removal generally have the properties shown in Table 3-4.

Consideration should be given to the typical operating conditions shown in Table 3-5.

Non-regenerable resins are not suitable for hardness removal. Raw water hardness concentrations are relatively high compared to other contaminants and would result in rapid exhaustion of the ion exchange resin. Resins would need to be replaced too frequently, on the order of days, to be practical.

Table 3-4 WAC resin typical physical and chemical characteristics

Characteristics	Description
Polymer structure	Porous cross-linked polyacrylic
Physical form	Spherical beads
Functional groups	Carboxylic acid
Ionic form, as shipped	H^+
Total capacity, Na^+ form	4.5 eq/L (98.3 kgr/ft^3) min
Moisture retention, Na^+ form	45–55%
Particle size range	300–600m; 1% max. <300m
Reversible swelling	Normal exhaustion 7% (approx.), full conversion 20% max.
Reversible swelling	70% (max.)
Specific gravity, Na^+ form	Approx. 1.19
Shipping weight, H^+ form	740–780 g/L (46.3–48.8 lb/ft^3)
Maximum temperature limit	120°C (250°F)

Table 3-5 WAC typical operating conditions for hardness removal

Step	Design Basis	Duration
Backwash	Set for minimum water temperature to give 50% bed expansion	1 BV on clean water supplies and 2–3 BV where solids are present
Bed settle	To allow the bed to reform fully classified	5–8 min
Acid injection	Typically 5–15% excess acid compared to the expected resin capacity. For hydrochloric acid regeneration, this is applied at approximately 4% HCl concentration at 2–4 BV/h (0.25–0.5 gpm/ft^3). The volume of regenerant must exceed resin volume. For sulfuric acid regeneration, this is applied at 0.5–1% at 8-20 BV/h (1–2.5 gpm/ft^3). 0.7% at 12–16 BV/h (1.5–2 gpm/ft^3) is normally encountered in operating plants.	Typically 20–45 min, depending on regeneration level and flow rate
Slow rinse	1–2 BV (0.25–0.5 gal/ft^3) at approximately regenerant flow rate.	Typically 30–40 min depending on regeneration level and flow rate
Final rinse	3–5 BV (22.5–45 gal/ft^3) preferably at service flow rate or alternatively 8–16 BV/h (60–120 gpm/ft^3)	Typically 10–20 min. The less slow rinse employed to more final rinse required.

BV: bed volume, BV/h: bed volume per hour

3.5 Bench and Pilot Testing

Chapter 13 of this handbook discusses the general process of piloting an ion exchange system. This section focuses on the specific considerations for hardness removal. As with all bench and pilot testing, the test should mimic and be scalable to the anticipated full-scale system. Bench-scale testing can help determine the optimal empty bed contact time (EBCT), screen resins to identify best performing options, and demonstrate treatment performance through multiple loading and regeneration cycles.

A comprehensive water quality analysis should be performed that includes all of the anions and cations. Chapter 2 presents a table describing the typical water quality analyses. In some locations, radium can be found with calcium and magnesium. Even if the combined radium 226 and 228 is below the MCL of 5 pCi/L, ion exchange resins will remove radium and the resin could potentially become a radioactive waste. The design engineer should provide the resin supplier with a comprehensive water quality analysis as described in Chapter 2 to determine the potential impacts of radionuclides.

Ion exchange has been used for hardness removal for more than 70 years, and the resin manufacturers have robust computer models that can be used for preliminary sizing. A pilot study is especially beneficial when multiple contaminants/ions require removal.

3.6 Design Guidelines

Recommended design guidelines for SAC and WAC regenerable resins for hardness removal are described below.

Pretreatment:

- Prefilters to prevent total suspended solids (TSS) accumulation in the resin bed is vital to satisfactory operation. The resin will act as an absolute filter producing water with an effluent turbidity of less than 0.3 NTU. A particle size distribution analysis can help determine the size of incoming solids. If the well produces any sand, a de-sander is also recommended. The feed turbidity to the system should be less than 1.0 NTU.
- Resin performs best when foulants are eliminated. If any of these are present in the water in concentrations high enough to affect resin performance, they need to be adequately removed:
  - Iron and manganese >1 ppm
  - Oil and grease—any concentration is detrimental
  - Oxidants—any concentration is detrimental, <0.2 ppm free chlorine for SAC resin
  - TOC >2 ppm

The recommended design guidelines for hardness removal are as follows.

SAC design parameters:

- Resin: strong acid cation in the sodium form
- Resin depth: 3 ft minimum (0.91 m)
- Hydraulic loading rate: 6–12 gpm/ft^2 (15–30 m/h)
- Specific flow rate: 1–5 gpm/ft^3 (8–40 BV/h)
- Empty bed contact time: 1.5–7.5 min minimum
- Vessels are generally designed for single vessel operation or parallel operation with at least one in regeneration or standby

General regeneration steps for SAC resin:

- Backwash (as needed): Use manufacturer's backwash curves to achieve 50 percent bed expansion for anywhere from 1–3 BVs depending on solids loading
- Regeneration using 10–12 percent sodium chloride solution at dosages ranging from 4 to 10 lb NaCl /ft^3 at a rate of 0.25–0.5 gpm/ft^3 (2–4 BV/h)
- Slow rinse step using raw water at 0.25–0.5 gpm/ft^3 (2–4 BV/h) for 30–40 min
- Fast rinse step using raw water at the service flow rate for 10–20 min

WAC design parameters:

- Resin: weak acid cation
- Resin depth: 3 ft minimum (0.91 m)
- Hydraulic loading rate: 6–12 gpm/ft^2 (15–30 m/h)
- Specific flow rate: 1–5 gpm/ft^3 (8–40 BV/h)
- Empty bed contact time: 1.5–7.5 min minimum

General regeneration steps for WAC resin:

- Backwash (as needed): Use manufacturer's backwash curves to achieve 50 percent bed expansion for anywhere from 1–3 BVs depending on solids loading
    - Regeneration using 0.5–4 percent hydrochloric acid (HCl) at dosage of approximately 200 g/L or 12.5 lb HCl/ft^3 of resin at a rate of 0.25–0.5 gpm/ft^3 (2–4 BV/h)
    - Slow rinse step with raw water at 0.25–0.5 gpm/ft^3 (2–4 BV/h) for 30–40 min
    - Fast rinse with raw water at the service flow rate for 10–20 min.

A key component in the design of ion exchange systems for hardness removal is the estimation of the amount of water that can be treated before the resin is exhausted (referred to as throughput). The throughput is based on the water quality parameters and is generally obtained

- by initial estimate of the capacity of the resin for the calcium and magnesium hardness; however, the capacity of the resin will be impacted by the additional contaminants in the raw water;
- by evaluation of similar systems;
- from estimation models used by resin suppliers; and
- from pilot testing.

3.7 Disposal Options

Hardness is generally present in concentrations that exhaust the resin relatively quickly, which results in many regeneration cycles and significant amounts of regenerant waste. Typically, the most preferred option is to send waste to the local public treatment works (PTW). If that is not an option, the waste may need to be hauled away by truck. To determine the most feasible option, the following will need to be considered:

- Waste volume
- Waste frequency
- Capacity to sewer
- Water quality requirements of the sewer

In some cases, equalization can be used to attenuate the flow if sewer capacity is an issue.

3.8 Design Example

A design example is provided for SAC.

3.8.1 SAC Example

In this example, a utility has drilled a new well with a capacity of 750 gpm to supplement its existing supply, and the new well was found to have a total hardness of 200 mg/L as $CaCO_3$. The utility decided to remove hardness to less than 5 ppm as $CaCO_3$ to make it available to blend with other wells to create a softened water source of 60 ppm as $CaCO_3$. A water analysis identified the composition of the well, which was used to design a softening plant. The water quality is summarized in Table 3-6. Note the water analysis is balanced for calcium carbonate equivalency.

Table 3-6 Balanced raw water chemistry

Raw Water Parameter	Value
Well pump flow rate	750 gpm
Firm treatment capacity	750 gpm
Sewer availability	Yes
Nitrate	12.0 mg/L as NO_3
Sulfate	96.15 mg/L
Chloride	106.4 mg/L
Bicarbonate	50 mg/L as $CaCO_3$
Calcium	60 mg/L
Magnesium	12.2 mg/L
Sodium	49.9 mg/L
Iron	0.2 mg/L
Manganese	0.05 mg/L
Total hardness	200 mg/L as $CaCO_3$
TOC	1 mg/L
Suspended solids	<0.2 mg/L
Conductivity	500 S/cm
pH	7.7

Table 3-7 Design parameters

Design Parameters	Value
Total flow rate	750 gpm
Influent hardness	200 mg/L as $CaCO_3$
Effluent concentration goal	5 ppb
Number of vessels in service	1
Number of vessels in standby	1
Number of vessels total	2
Flow per vessel	750 gpm
Vessel diameter	9 feet
Volume per supersack	42 ft^3
Number of supersacks per vessel	5
Media volume per vessel	210 ft^3
Vessel area	63.6 ft^2
Bed depth per vessel	3.3 feet
Linear velocity: 8–12 gpm/ft^2 target per vessel	12 gpm/ft^2
Specific flow rate: 1–5 gpm/ft^3 target per vessel	3.57 gpm/ft^3
EBCT	2.1 minutes
BV/hour	28.5
Estimated throughput, BV (provided by resin supplier)	500 bed volumes
Bed life	17.5 hours
Backwash linear velocity	2.4 gpm/ft^2

BV: bed volume, EBCT: empty bed contact time

While models are used to develop a plant design and performance, an estimate can be made during a preliminary discussion with the utility using the following parameters and estimated values in Table 3-7.

1. Choose the smallest commercially available standard diameter vessel that will provide sufficient cross-sectional area to produce a linear velocity in the target range of 8–12 gpm/ft^2. Table 3-8 presents commercially available vessel diameter with the cross-sectional area.

Determine available equipment diameter using the linear velocity criteria:

- Assume linear velocity: 12 gpm/ft^2
- Vessel area = treated flow / linear velocity

$$\frac{750 \text{ gpm}}{12 \text{ gpm/ft}^2} = 62.5 \text{ ft}^2$$

Table 3-8 Commercially available vessel diameters with cross-sectional area

Commercial Vessel Diameter (ID) ft	Approx. Cross-sectional area (ft^2)
12	113
11	95
10	78.5
9	63.6
8	50
7	38.5
6	28.3
5	19.6
4	12.6

In this case, a 9 ft diameter vessel will be chosen.

2. Calculate resin volume.

<u>Co-flow Regeneration</u>
For co-flow regeneration, a minimum resin bed depth of approximately 3 ft is recommended. For a 9 ft diameter vessel, the minimum resin volume needed is as follows:

- Cross-sectional area of 63.6 ft^2 × 3 ft bed depth = 190 ft^3 approx.
- Since resin deliveries are usually in supersacks of 42 ft^3 each, the nearest whole number of supersacks should be chosen: 5 supersacks × 42 ft^3 = 210 ft^3
- Recheck resin bed depth: 210 ft^3 / 63.6 ft^2 = 3.3 ft—this is acceptable.
- Determine the linear velocity:

$$750 \text{ gpm} \div 63.6 \text{ ft}^2 = 11.8 \text{ gpm/ft}^2$$

- 12 gpm/ft^2 is in the acceptable range of 8 to 12 gpm/ft^2 target per vessel.

Determine the specific flow rate:

$$750 \text{ gpm} \div 210 \text{ ft}^3 = 3.57 \text{ gpm/ft}^3$$

- 3.6 gpm/ft^3 is in the acceptable range of 1 to 5 gpm/ft^3 target per vessel.

Determine EBCT:
EBCT = vessel volume / treated flow rate

$$210\ \text{ft}^3 \times \frac{7.48\ \text{gal}}{\text{ft}^3} \times \frac{\text{min}}{750\ \text{gal}} = 2.1\ \text{min}$$

- 2.1 min is within the EBCT goal criteria.

Estimated Throughput

The resin supplier estimates that the resin can treat 500 BV of water before regeneration is required.

$$500\ \text{BV} \times \frac{210\ \text{ft}^3}{\text{BV}} \times \frac{7.48\ \text{gal}}{\text{ft}^3} \times \frac{1\ \text{min}}{750\ \text{gal}} = 1{,}047\ \text{min} = 17.5\ \text{h}$$

- The resin will require regeneration after approximately 17.5 h of run time.

Estimate throughput rate:

$$\frac{750\ \text{gal}}{\text{min}} \times \frac{1\ \text{ft}^3}{7.48\ \text{gal}} \times \frac{60\ \text{min}}{1\ \text{h}} \times \frac{1\ \text{BV}}{210\ \text{ft}^3} = 28.6\ \text{BV/h}$$

Design Regeneration Protocol

Regeneration steps:

- Backwash—typically expands the resin bed by 50–60 percent for 10 min (see note that follows)
- Regeneration—either co-flow or counter-flow
- Slow rinse—typically 7.5–15 gal/ft^3 of resin (1–2 BV)
- Fast rinse—typically 23–38 gal/ft^3 of resin (3–5 BV)

Backwash Note: The backwashing step is done as part of each regeneration procedure if the system is designed for co-flow regeneration. For counter-flow regeneration, backwashing may be done infrequently, on the order of once per month. Counter-flow backwashing has the potential to cause unintentional mixing of the highly regenerated resin at the outlet of the resin bed with the rest of the resin that is less so. This can lead to potentially higher than normal nitrate leakage during the following cycle. When backwashing is eventually done, the resin bed would usually be double regenerated before returning to service. During backwashing, the resin bed is generally expanded by 50–60 percent of its original settled volume to remove as much of the suspended solids that may have accumulated on the resin beads during the previous service period. It is important to get advice from the resin supplier on relevant backwash linear velocity rates that will accomplish the desired degree of bed expansion at the specific temperature of the water. A typical backwash rate at 50°F (10°C) can be 2 gpm/ft^2 (5 m/h). For a 9 ft (2.74 m) diameter vessel with a cross-sectional area of 63.3 ft^2, that equates to a flow rate of 126.6 gpm (28.75 m^3/h). Usually, backwashing is done for at least 10–15 min;

if the backwash effluent water is not relatively clear of suspended solids after this period, the backwash can be continued until effluent quality is satisfactory.

Brine Makeup and Application: A saturated brine solution (typically 26 percent) is usually prepared using sodium chloride salt that meets acceptable quality standards (e.g., ANSI/AWWA B200-17). Saturated brine contains approximately 2.6 lb of salt per gallon (approximately 300 g/L). During regeneration, the saturated brine solution is mixed with enough dilution water to create the desired concentration of brine (typically 5–12 percent) before it is passed through the resin bed. This requires a dilution ratio of roughly 2:1 for dilution water vs saturated brine. At 10 percent brine concentration, the salt content of each gallon of 10 percent brine will be approximately 0.9 lb.

Volume and Flowrate of 10 Percent Brine Solution: In this example, assume that the salt dosage applied is 10 lb salt per 1 ft^3 of resin (or 160 g/L). Using a volume of 210 ft^3 of resin in co-flow regeneration mode, the total volume of 10 percent brine needed can be calculated from:

$$210 \text{ ft}^3 \text{ of resin} \times \frac{10 \text{ lb of salt}}{\text{ft}^3 \text{ of resin}} = 2{,}100 \text{ lb of salt}$$

Therefore, the volume of 10 percent brine needed is:

$$\text{Vol of 10\% brine} = 2{,}100 \text{ lb of salt} \times \frac{1 \text{ gal}}{0.9 \text{ lb of salt}} = 2{,}333 \text{ gal of 10\% salt}$$

Brine is usually applied at a specific flow rate of 0.25–0.5 gpm/ft^3 (2–4 BV/h)—the more contact time, the better the efficiency. Ideally, the contact time between the brine and the resin should be a minimum of 30 min.

At 0.25 gpm/ft^3 and a volume of 2,333 gal of 10 percent brine, the brining time will be:

$$\frac{2{,}333 \text{ gal}}{210 \text{ ft}^3} \times \frac{\text{ft}^3}{0.25 \text{ gpm}} = 45 \text{ min(approx.)} \rightarrow \frac{2{,}333 \text{ gal}}{45 \text{ min}} = 53 \text{ gpm}$$

Following the brine, the resin is usually rinsed to rid it of residual brine. Usually, the resin must be rinsed until the chloride content of the rinse water is about 25 ppm or lower compared to the chloride level in the raw water. At that point, the resin bed can be put into service. The first phase of rinsing, known as the slow rinse step, is conducted at the same flow rate used during the brining step. The objective of this is to continue the regeneration process of the resin by pushing out the brine in plug-flow fashion and ensuring that the nitrate and other anions are adequately removed from the resin in preparation for the next cycle.

For the slow rinse, 7.5–15 gal of water per cubic feet of resin is typically used.

Slow rinse water usage:

$$210\ \text{ft}^3\ \text{of resin} \times \frac{15\ \text{gal of water}}{\text{ft}^3\ \text{of resin}} = 3{,}150\ \text{gal of water}$$

Using the same 0.25 gpm/ft³ specific flow rate used for the brining step, the slow rinse will be for 60 min and the flow rate will be 63 gpm.

$$\frac{3{,}150\ \text{gal}}{210\ \text{ft}^3} \times \frac{\text{ft}^3}{0.25\ \text{gpm}} = 60\ \text{min} \rightarrow \frac{3{,}150\ \text{gal}}{60\ \text{min}} = 53\ \text{gpm}$$

For the fast rinse step, the service flow rate can be used to save time. In this case, the service flow rate is 750 gpm, so the estimated time for the fast rinse, assuming 38 gal/ft³ (5 BV) will be approximately:

$$210\ \text{ft}^3 \times \frac{38\ \text{gal}}{\text{ft}^3} \div 750\ \text{gpm} = 11\ \text{min (approx.)}$$

The previous estimates are summarized in Table 3-9.

Figure 3-1 is an illustration of an example SBA regeneration co-flow regeneration.

Residuals handling: Total wastewater volume expected can be computed from the addition of the volumes of backwash, brine, and rinse. In this case, the wastewater volume is approximately 17,620 gal (about 9 BV) or about 1.8 percent of the water treated. The composition of the spent brine will be a mixture of sodium salts comprised largely of chloride but also sulfate, nitrate, and bicarbonate that was eluted from the resin.

Table 3-9 Regeneration based on 10 lb/ft³ NaCl and 10% brine solution

	Design Basis	Flow (gpm)	Time (min)	Total Waste per Step (gallons)
Backwash	2 gpm/ft²	127	10	1,270
Brine step	0.25 gpm/ft³	53	45	2,385
Slow rinse	0.25 gpm/ft³	53	60	3,180
Fast rinse		750	11	8,250
Totals			126	15,085

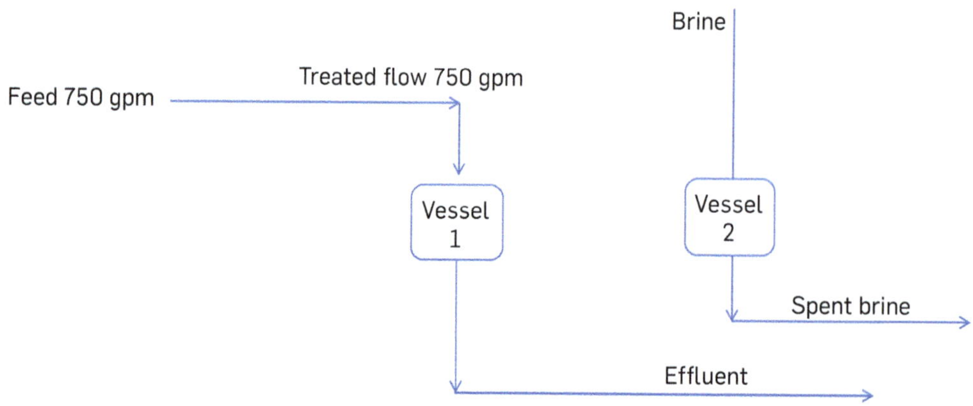

Figure 3-1 SBA regeneration co-flow regeneration example: vessel 2 is off-line while in regeneration

Courtesy of Purolite Corp.

REFERENCES

Edzwald, J.K. 2011. *Water Quality and Treatment: A Handbook on Drinking Water*. 6th ed. AWWA, American Society of Civil Engineers, McGraw-Hill. https://www.awwa.org/Store/{ProductName}/ProductDetail/{ProductId}?productId=6292.

Pingale, D.K. 2005. "Role of Weak Acid Cation Resin in Water Treatment." *Chemical Weekly*, February.

WHO (World Health Organization). 2011. "Hardness in Drinking-water." 10.01. https://www.who.int/water_sanitation_health/dwq/chemicals/hardness.pdf.

CHAPTER 4

Total Organic Carbon Removal

4.1 Introduction

Organic matter in surface and groundwater is a diverse mixture of organic compounds ranging from macromolecules to low molecular weight compounds. Organic carbon is typically measured as total organic carbon (TOC) and dissolved organic carbon (DOC), which is generally taken as that which will pass through a 0.45 μm filter. Organic carbon in surface and groundwater sources does not directly cause adverse health effects, but rather is a precursor to disinfectant byproducts (DBPs). Concentrations of TOC and DOC generally above 1–2 mg/L can have the following impacts:

- Cause the formation of disinfection byproducts including total trihalomethanes and haloacetic acids
- Increase taste and odors
- Increase oxidant demand
- Impact other treatment technologies, including
  - Foul membrane systems
  - Foul UV lamps and decrease the effectiveness of UV disinfection
  - Foul media like ion exchange
- Cause an increase in biofilm growth in distribution systems

This chapter describes how ion exchange in both fluidized and fixed bed forms can be used for organic carbon removal from surface and groundwater sources in order to provide safe drinking water that satisfies the regulatory objectives. Although regulations are written in terms of TOC, ion exchange is primarily effective in removing DOC. There are resins that remove both dissolved and colloidal TOC. There are highly microporous resins that function both by the ion exchange and adsorption mechanisms and which contain pores large enough to trap colloidal organics. Depending on the water source and other water quality parameters DOC is typically between 90 and 99 percent of the TOC in the water. This chapter describes:

- properties and chemistry associated with the forms of organic carbon found in groundwater and surface water sources;
- regulations regarding organic carbon: specifically, the US Environmental Agency (USEPA), the World Health Organization (WHO), United Kingdom (UK) Water Supply Regulations, European Union Directive, and recently enacted Scottish Water organic carbon goals;
- source water quality parameters that can impact the selection of ion exchange resins suitable for organic carbon removal;
- an overview of the ion exchange configurations and types of resins that can be used for organic carbon removal;
- an overview of bench and pilot testing requirements;
- design guidelines including calculations and residuals generation; and
- a design example.

4.2 Organic Carbon Properties and Occurrence

4.2.1 Chemistry

Organic carbon has significantly different properties than inorganic carbon. Inorganic carbon exists in water in the form of carbon dioxide (CO_2), carbonate (CO_3^{-2}), bicarbonate (HCO_3^-), carbonic acid (H_2CO_3). Inorganic carbon impacts the removal of hardness and is the primary component of alkalinity in water.

Organic carbon compounds in water are characterized by at least two carbon atoms and are covalently bonded with each other ("Organic Compounds Definition, Examples, List, And Structure" n.d.). For drinking water systems, TOC is typically used for defining the quantity of organic carbon in the water and is defined as the total concentration of organic carbon in the water. TOC is composed of DOC and particulate organic carbon (POC). DOC is determined by filtering the water sample through a 0.45 µm filter and then measuring the organic carbon concentration. POC is determined by subtracting the DOC from the TOC. POC can generally be removed through conventional filtration. This chapter focuses on the dissolved portion (DOC) which anion exchange resins are able to remove.

Individual organic carbon species are typically not measured in raw water sources but often the organic carbon found in water has a negative charge. This negative charge allows for the use of anion exchange for removal of a portion of the organic carbon.

The surface charge of particles in surface water becomes less negative with decreasing pH. These charged particles repel each other, allowing them to remain suspended in water. The charge of these particles also depends on the concentration of positively charged ions in the waters (Ca^{2+}, Mg^{2+}, etc.), the composition (minerals, organic detritus, living cells, etc.), and the concentration of DOC in the water (Peternelj 2009).

Organic carbon compounds can be further classified into humic and non-humic compounds. Humic compounds are derived from the decay of plant and animal matter. Non-humic substances are produced from plant and algal byproducts. The proportion of humic and non-humic substances will impact the effectiveness of traditional coagulation for removal of organic carbon and is discussed in Section 4.3. Humic substances are more amenable to removal by coagulation than non-humic matter. Specific ultraviolet absorbance (SUVA) is a metric that uses UV_{254} and DOC to assess the relative proportion of humic and non-humic substances. SUVA is defined as:

$$\text{SUVA} = \frac{UV_{254}\left(cm^{-1}\right) \times 100}{DOC(mg/L)} \qquad \text{(Eq 4-1)}$$

Waters with a SUVA greater than 4 have organic matter that is dominated by humic substances. A SUVA between 2 and 4 has a mix of humic and non-humic substances. A SUVA of less than 2 is dominated by non-humic substances (Edzwald 2011). Waters with SUVAs below 4 are well suited for consideration for removal by ion exchange.

Hydrophobicity also plays a role in whether coagulation or anion exchange is a more efficient removal mechanism. Anion exchange resin can remove a wider range of hydrophilic, transphilic, and hydrophobic fractions and molecular weight fractions than coagulation or activated carbon adsorption (Boyer 2015).

4.2.2 Occurrence

Unless the raw water source is associated with some form of man-made contamination, organic carbon in surface and groundwaters is derived from the decay of plant and animal matter. In both groundwater and surface water, carbon is primarily found in the dissolved rather than particulate form, accounting for approximately 90 to 99 percent of TOC in groundwaters and typically 100 percent of TOC in surface water (Edzwald 2011). TOC in surface water sources will vary seasonally and some sources have been experiencing an increase over the past 20 years. According to an article in BioScience, some surface waters have experienced an increase in TOC of 39 percent from 1999 to 2009. The increase in TOC could be due in part to increased algal blooms caused by climate change (Erlandsson et al. 2011).

4.3 TOC and Disinfection Byproduct Regulations

4.3.1 USEPA Regulations and Standards

In December 1998, USEPA promulgated the Stage 1 Disinfection Byproducts (DBP) Rule. The purpose of the DBP Rule is to reduce exposure to disinfectant

Table 4-1 TOC removal requirements

Source Water TOC (mg/L)	Source Water Alkalinity (mg/L as $CaCO_3$)		
	0–60	>60 –120	>120
>2.0–4.0	35.0%	25.0%	15.0%
>4.0–8.0	45.0%	35.0%	25.0%
>8.0	50.0%	40.0%	30.0%

byproducts to the consumer. The primary precursor to DBP formation is natural organic matter, typically measured as TOC. Systems using conventional treatment must remove specific amounts of TOC before adding disinfectants. The required percent removal depends on the source water TOC and alkalinity, as outlined in Table 4-1. Table 4-1 presents the TOC removal requirements.

As discussed previously, organic carbon by itself does not cause adverse health consequences; these are generally the result of DBPs formed during the reaction between chemical disinfectants and organic matter. Total trihalomethanes (TTHM) and haloacetic acids (HAA) are common DBPs that are regulated with a maximum contaminant level (MCL) under the DBP Rules. TTHM is the sum of four trihalomethanes and has an MCL of 80 µg/L based on a locational running annual average under the DBP Rule. HAA5 is the sum of five regulated haloacetic acids with an MCL of 60 µg/L based on a locational running annual average. The fourth Unregulated Contaminant Monitoring Rule (UCMR4) requires monitoring of four additional HAAs for a total of nine HAAs (HAA9). Utilities are required to take samples at the sample locations as the HAA5 samples and report results. Bromate is another disinfection byproduct that is being found with increasing frequency due to increased bromide in raw waters. Bromate has an MCL of 10 µg/L. Raw water bromide is increasing as a result of scrubbing processes at coal-fired power plants, which generate a bromide waste that is discharged to water bodies. Sea water intrusion at water treatment plants located near the ocean is also becoming a problem. Bromide is an anion and can also be removed by ion exchange.

The TOC in some waters is not amenable to removal by coagulation and USEPA established an alternative compliance criteria that allows exemption of the TOC removal requirement if the raw or treated water SUVA is less than two. For systems that cannot meet the TOC removal requirements but have elevated disinfection byproducts, anion exchange may be an appropriate treatment technology for removal of organic carbon.

4.3.2 International Regulations and Guidelines

WHO, UK Water Supply Regulations, and European Union European Commission Directives have not established any TOC levels or removal requirements. Rather, they have established guidelines and regulations for select DBPs: TTHM, HAA, and bromate. Table 4-2 provides a summary of these guidelines/regulations.

Table 4-2 Regulation summary

Disinfection Byproducts	WHO Guidelines 4th ed. 2011	UK Water Supply (Water Quality) Regulations Prescribed Concentration	EC Directive 98/83/EC November 1998 Parametric Concentration	USEPA Primary Drinking Water Regulations MCL
Bromate (BrO_3)	10 µg/L*	10 µg/L	10 µg/L	10 µg/L
Haloacetic acids				60 µg/L as a total of 5 acids
Monochloroacetate	20 µg/L			
Dichloroacetate	50 µg/L*			
Trichloroacetate	200 µg/L			
Monobromoacete				
Dibromoacetate				
Chlorophenols				
2,4,6-Trichlorophenol	200 µg/L			
Halogenated acetonitriles				
Dichloroacetonitrile	20 µg/L*			
Dibromoacetonitrile	70 µg/L			
Trihalomethanes				
Bromoform	100 µg/L†	100 µg/L sum of concentrations	100 µg/L sum of concentrations	80 µg/L sum of concentrations‡
Chloroform	300 µg/L†			
Dibromochloromethane	100 µg/L†			
Bromodichloromethane	60 µg/L†			

* Provisional guideline
† The sum of the ratios of concentration for each to their respective GV not to exceed 1
‡ Calculated based on the Locational Running Annual Average (LRAA)

4.4 Ion Exchange Resin Treatment Options

The USEPA Drinking Water Treatability database identifies the following treatment technologies for the removal of organic carbon:

- Conventional treatment coagulants in conjunction with clarification and filtration
- Biological filtration
- Granular activated carbon
- Ion exchange
- Membrane filtration using ultrafiltration and nanofiltration

Ion exchange for DOC removal currently has two different configuration options: fixed bed or fluidized bed systems. Fluidization is a process where granular material (resin beads) are induced to behave as a fluid by passing a fluid (liquid or gas) through a bed of the granular material. By moving the media along with the flow of water, less pressure is required to drive the process, which may be an advantage at high flow rates. Conventionally, ion exchange operates in a fixed bed where the water passes down through a column of resin under pressure. Fluidized bed systems, largely used in the mining industry, were introduced to the drinking water industry approximately 20 years ago.

Fluidized ion exchange systems offer the benefit of treating water largely independent of total suspended solids (TSS). Since there is no need for the removal of particulates or solids, fluidized systems are typically at the influent of a plant. These systems are well suited for surface water treatment as surface waters may have high TSS. For surface water systems, ion exchange is generally suitable when:

- conventional treatment is insufficient to achieve the TOC removal requirements;
- when the coagulant concentrations required to achieve the necessary TOC removal are very high, resulting in high coagulant cost and high sludge treatment and disposal costs; or
- when additional TOC removal is needed due to elevated DBPs in the distribution system.

For groundwater systems, ion exchange can be a viable option when there is elevated DOC resulting in high DBPs in the distribution system. Given that groundwater sources typically have low TSS, fixed bed ion exchange may be a good option as pretreatment may not be not needed. Ion exchange in a pressurized configuration may eliminate the need for repumping. Figure 4-1 outlines the options and applications of ion exchange for DOC removal.

4.4.1.1 Ion Exchange Resin for DOC Removal

Both Type I and Type II strong base anion (SBA) exchange resins are effective at removing the dissolved portion of TOC. Often the DOC is weakly negatively charged and therefore Type I resins generally provide greater removal because they are more strongly basic. Type I resins generally have less DOC leakage than

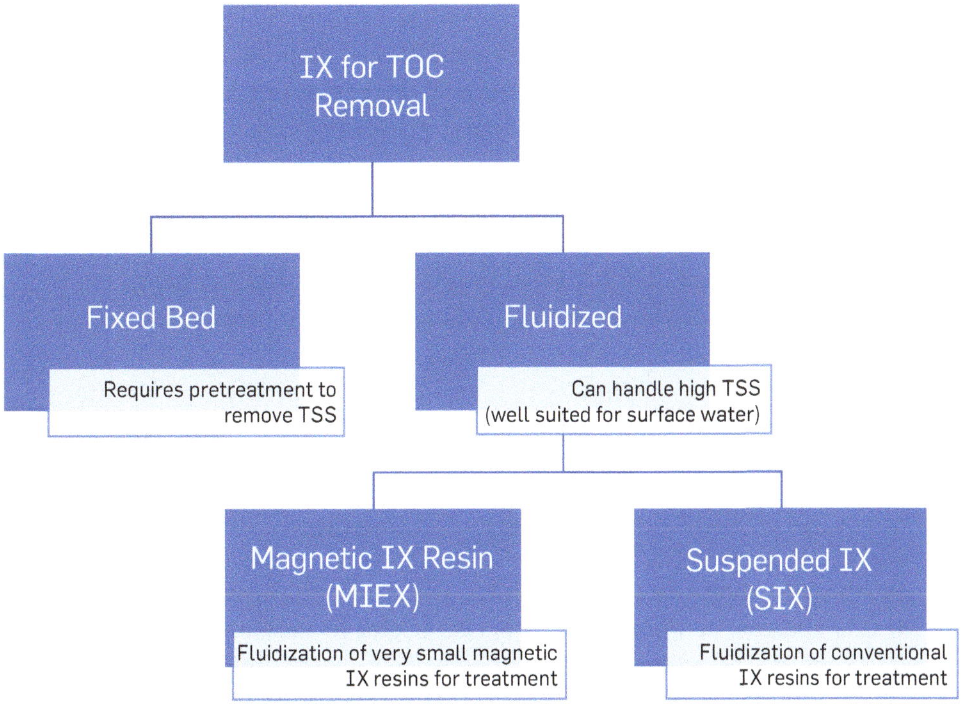

Figure 4-1 Ion exchange (IX) options for TOC removal

the other resin types (however, keep in mind that leakage is a function of applying resins in a column and do not apply to fluidized processes that maintain a continuous supply of regenerated resin back to the process). Type II resin has the benefit of experiencing less fouling than Type I (Galletti 2000). SBA exchange resins are effective over a wide pH range and do not typically require pH adjustment.

Macroporous resins with higher surface areas are preferred, as they have a strong attraction for the larger molecular weight aromatic portion of DOC and tend to resist fouling better than gel type resins. Acrylic type resins are generally less prone than styrenic resins to long-term fouling by high molecular weight fractions of DOC.

The binding capacity of a resin for the ionized fraction of DOC will depend both on its relative affinity for the DOC versus major anions present in the water (e.g., sulfate and chloride) and their relative concentrations. The concentration of competing anions will determine the loading rate and regeneration frequency.

Generally, the anion exchange reaction is

$$\text{DOC}^- + \text{Resin} - \text{Cl}_2 \leftarrow \rightarrow \text{Resin} - \text{DOC} + 2\text{Cl}^-$$

The reaction is largely reversible for lower molecular weight fractions of DOC. During regeneration, a brine solution is passed over the resin and chloride is exchanged for DOC and the resin can be put back into service. Chapter 2, Fundamentals of Ion Exchange, goes into greater detail on the ion exchange process and provides a graphical representation of these principles. This process can generate a large excess of chloride that may need special disposal considerations.

Ion exchange is capable of removing 50–80 percent of DOC depending on the nature of the DOC, background water chemistry and the selectivity of the chosen resin.

As with all water treatment applications, a detailed raw water analysis is needed for proper resin selection and system design. Water parameters needed include the ions outlined in Chapter 2, Fundamentals of Ion Exchange.

For resin selection, the following parameters should be measured

TOC	Iron	Total alkalinity	Nitrate
DOC	Manganese	Ca hardness	Sulfate
UVA	Turbidity	Total hardness	Bromide
pH	Conductivity	Chloride	Color

Analysis of competing anions (sulfate, chloride, etc.) is critical to determining loading and regeneration rates. It is also useful to measure the concentration of metals such as iron and manganese, which can cause scaling or form complexes with organic matter than can be more difficult to remove from the resin. These necessitate pretreatment and oxidant to maintain the resin. Fluidized beds operate independently of turbidity and suspended solids, but these must be considered for fixed column configurations.

DOC is a broad term that includes many different molecular weight species comprised of varying humic and non-humic substances. Some DOC is amenable to being reduced efficiently by ion exchange alone, while waters with similar levels of DOC would be better treated with coagulation, and others would require both coagulation and ion exchange to achieve similar DOC reductions. The best way to determine which approach or combination of approaches will produce the most DOC reduction is to perform a pilot study or jar tests for fluidized systems. The advantage of performing these tests is that they will allow the user to precisely predict not only the percent removal of DOC, but to scale up the process for capital and operational planning purposes.

4.4.2 Fixed Ion Exchange Systems

Fixed bed systems require influent water with very low to no TSS; otherwise, the system would become easily fouled and require frequent backwashing. They are well suited for groundwater treatment, which naturally has very low TSS. Fixed bed systems can also be implemented at a surface water treatment plant downstream of the gravity filters after TSS has been removed.

4.4.2.1 Fixed Bed Configuration

Using a regenerable SBA resin is feasible when the concentrations of sulfate and other competing anions are moderate and regeneration is needed every few weeks or longer. Some plants run beyond the sulfate breakthrough because the small TOC peaking that occurs at the sulfate breakthrough may still be acceptable, depending on the percent TOC reduction being targeted.

When a vessel of resin has reached DOC breakthrough, the vessel is taken out of service and regenerated. During the regeneration step, 1.4 to 2 bed volumes (BVs) of sodium chloride (NaCl) brine regenerant are pumped through the resin at concentrations of 8–12 lb of salt/ft^3 to strip the DOC and other anions from the resin and put the resin back in the chloride form. The brine waste will contain high concentrations of both sodium chloride and organic carbon. The system then goes through a slow rinse to displace the last contaminants followed by a fast rinse, mainly to displace the NaCl. The vessel is then ready to be placed back in service.

4.4.2.2 Fixed Bed Design Guidelines

Recommended design guidelines for fixed bed SBA resins are further described here.

Pretreatment:

- Prefilters to prevent TSS accumulation in the resin bed are vital to satisfactory operation.
- Resin performs best when foulants are eliminated. If any of these are present in the water in concentrations high enough to affect resin performance, they need to be adequately removed:
  - Turbidity <1 NTU
    - filtration with filters
  - Particulate iron and manganese >0.3 mg/L
  - Oil and grease—any concentration is detrimental
  - Oxidants—any concentration is detrimental
  - LSI—ensure there is no scaling potential, <0

SBA Design Parameters:

- Resin: SBA
- Resin depth: 3 ft minimum (0.91 m)
- Hydraulic loading rate: 8–12 gpm/ft^2 (20–30 m/h)
- Specific flow rate: 1–2.5 gpm/ft^3 (8–20 BVs/hour)
- Empty bed contact time: 3–7.5 min minimum
- Regenerant solution: Sodium chloride brine
- Regenerant concentration: 6–12 percent NaCl
- Regenerant dosage: approx. 8–10 lb NaCl/ft^3 of resin

Regeneration Steps:

- Backwash (as needed): Use manufacturer's backwash curves to achieve 50 percent bed expansion for anywhere from 1 to 3 BVs depending on solids loading
- Brine step with regenerant solution: 1–1.5 BV with 10 percent NaCl, about 8–10 lb NaCl/ft^3 of resin
- Slow rinse with treated water: typically 0.25–0.5 gpm/ft^3 (2–4 BV/h) for 30–40 min

- Fast rinse with treated water: typically at the service flow rate for 10–20 min

A key component in the design of ion exchange systems for DOC removal is the estimation of the amount of water that can be treated before the resin is exhausted (referred to as throughput). The throughput is based on the water quality parameters and is generally obtained

- by evaluation of similar systems,
- from the resin suppliers, each of which have models that allow an estimation of throughput, and
- from pilot testing.

4.4.2.3 Brine Disposal Options

The impacts of the spent brine on wastewater treatment plant discharges and National Pollutant Discharge Elimination System (NPDES) requirements (USEPA 2007) must be evaluated for each system. Sewer disposal, evaporation ponds, brine haul off, deep well injection, and disposal via a brine line are other options as well.

Spent brine will also concentrate other anions in the water. Complete characterization is necessary to ensure the uranium concentration and that the other ionic species are compatible with the requirements of the disposal site.

4.4.3 Fluidized Bed Ion Exchange Systems

Fluidized bed ion exchange systems offer the benefit of not requiring pretreatment to remove TSS. Conventional fixed bed ion exchange columns act as a filter and may be easily occluded, which is not the case when resin is mixed freely with water. Potential benefits of these alternative configurations include:

- much lower head loss (generally less than 24 in.),
- the option to treat without pumping (gravity or pressure vessel options),
- well suited for treatment of surface water,
- no chromatographic peaking (no leakage) due to the fact that regenerated resin may be continually returned to the treatment basin, and
- separate basins for containment of treated water and concentrated chloride regenerants for well head or cross contamination protection.

Fluidized ion exchange systems are also a good option if organic carbon pretreatment is needed ahead of another unit process.

These fluidized systems are generally suitable for surface waters under the following conditions:

- When conventional coagulation does not provide sufficient TOC removal

- When enhanced coagulation conditions are expensive, pH conditions produce unanticipated consequences of corrosion, and/or the residuals from these conditions are onerous and expensive
- When DBPs exceed the regulatory limit or goals of the utility even with enhanced coagulation
- When other options for TOC/DBP reduction are not cost effective
- When the raw water has elevated concentrations of other anions such as bromide or sulfate that can also be removed using anion exchange

4.4.3.1 Fluidized Bed Treatment Configuration

In contrast to fixed bed ion exchange systems where the resin is contained in a column and the water passes through the resin bed, fluidized systems have the resin mix with the water in a contact (treatment) basin. The resin is separated from the water and reactivated in a sidestream process and then reenters the contact basin. This means that the water treatment part of the cycle operates at steady state loading.

For both fixed bed and fluidized bed processes, bed volume treatment rate (BVTR) is the ratio of water to resin. With fixed bed ion exchange, the treatment performance and effluent water quality gradually decline throughout a treatment cycle until a regeneration is initiated. However, for the fluidized process, external regenerations allow for the semi-continuous return of regenerated resin to the process to keep the BVTR, and therefore the treated water quality, stable.

Resin regeneration is conducted separately from the main water treatment process. Small volumes of the total resin in the contact basin are periodically removed to external tanks to be regenerated with a sodium chloride solution and then returned to the main contactor to maintain a consistent ion exchange capacity. The withdrawal and return of resin ensure consistent treated water quality and prevents chromatographic peaking that can occur with conventional fixed bed ion exchange vessels.

A programable logic controller (PLC) controls the timing of the regeneration process based on actual flows and seasonal DOC fluctuations (often correlated to UVA for online process control); therefore, if the regeneration vessels are sized for worst-case conditions, only the actual volume of resin required to achieve efficacy needs to be regenerated each day. Plug flow conditions in the desorption (regeneration) process (versus in the sorption or treatment process) offer the efficiencies of plug flow conditions without the disadvantages that come from treating in a plug flow column (primarily the limitation of suspended solids, and chromatographic peaking/leakage). Specific design features of the regeneration vessel are dependent on the type/kind of resin used, dewatering methods required, transport mechanisms used, and the efficiency of the plug flow vessel itself. Redundancy should be considered in the design of the regeneration vessels as these are critical to maintaining adequate DOC reduction and will operate under plug flow conditions for maximum efficiency of desorption kinetics—which means that regeneration vessels will be anion exchange columns and by default will have periods in which they are online/offline.

The contact (treatment) basin is operated to maintain optimal mixing of the resin/water up to the peak loading rates of the basin design. POC, suspended solids, and other floating solids will pass through the fluidized bed treatment basin and on to the next unit process (oftentimes coagulation/flocculation/sedimentation/filtration). Split flow (allowing a portion of the raw water to bypass pretreatment) is an option for fluidized bed processes.

As the resin itself is moving throughout the treatment system, there is an opportunity for resin to be lost as it moves from the contact basin through the regeneration process and back to the contact basin. Virgin resin must be dosed to maintain the desired resin concentration in the contact basin. The cost of new resin should be considered in the design process.

The brine from ion exchange systems (both fluidized and suspended) is a function of the volume of resin regenerated per day, the concentration of salt brine used for regeneration (dependent on the type and kind of resin), brine reuse strategies, and rinsing requirements to reduce chloride carryover with the regenerated resin.

4.4.3.2 Fluidized Bed System Manufacturers

Currently, there are at least two manufacturers of fluidized ion exchange systems: IXOM, developer of the MIEX process, and PWNT (Velserbroek, Netherlands), developer of the suspended ion exchange (SIX) process. The MIEX process uses small bead magnetic ion exchange resins, while the SIX process uses conventional large bead resins.

MIEX stands for magnetic ion exchange. MIEX uses SBA Type 1 resins with a magnetized iron oxide backbone that are manufactured for use in MIEX processes by IXOM (the process developer is the manufacturer of the resin). The resin beads are extremely small—often a fraction of the size of typical conventional anion exchange resins. Kinetics are inversely proportional to particle size; therefore, ideal kinetics (faster exchange, smallest footprint) would come from the smallest practical bead size. Their small bead size provides a large surface area enabling rapid ion exchange kinetics in the adsorption and desorption mode. The magnetism does not aid in DOC removal but rather results in the beads forming agglomerates that settle rapidly or fluidize at high hydraulic loading rates. It is also used as a mechanism to control and retrieve the extremely small resin beads. The MIEX process was developed in 1990, and there are approximately 80 drinking water facilities worldwide and 55 in the United States using the MIEX system. MIEX resin is tolerant of oxidants (up to limitations established by IXOM); the process is tolerant of heavy metals (the limitation is non–scale-forming waters) and can tolerate continual use of saturated salt (or sodium bicarbonate) as regenerants. There are no cases of MIEX anion exchange resins being fouled by the likes of high DOC waters (found in the uplands of the United Kingdom and in some Florida surface waters), largely because these resins are not exhausted before regeneration, are regularly exposed to saturated solutions of regenerate that completely desorb DOC, and have rapid exchange kinetics. MIEX resins were developed primarily for removal of DOC from municipal drinking water plants.

The MIEX process currently has three different delivery methods, each with a unique treatment basin design based on treatment flow, concentration and contact time (CT) of resin, and limitations of site and hydraulic profile:

- High Rate (200 ml/L of resin and 3–6 min CT)
- Dual Stage (20 ml/L of resin and 10–20 min CT)
- Integra—resin concentration and CT dependent on existing basin geometries. The Integra system is a trade name from IXOM and is customized for the application.

Each process uses a different concentration of resin depending on specifics such as basin geometry and hydraulic loading rates. All MIEX systems use the small bead MIEX resins for uniform dispersion, fast kinetics, easy dewatering, and ease of resin transfer. Figure 4-2 provides an example process flow diagram for the MIEX High Rate method.

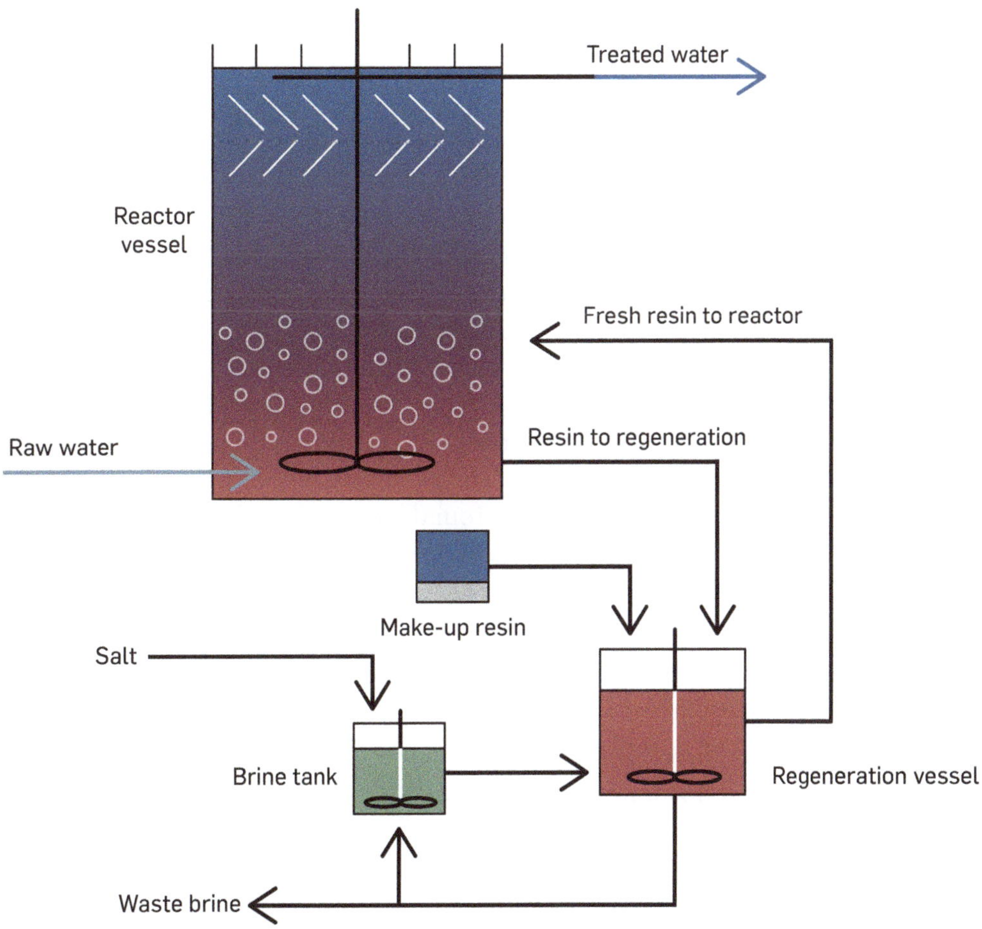

Figure 4-2 MIEX high rate process flow diagram

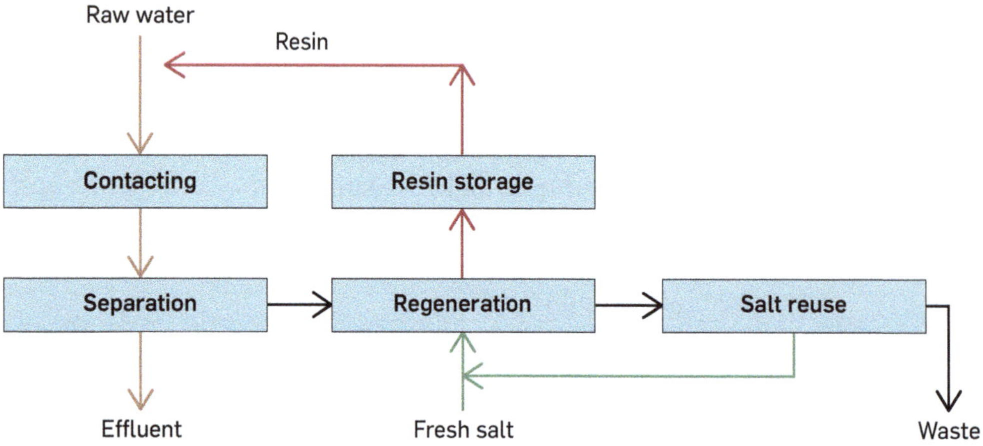

Figure 4-3 Schematic of SIX process of PWNT

The second fluidized ion exchange process is referred to as the SIX process and was developed by PWNT. SIX stands for suspended ion exchange. SIX is an advanced plug flow reactor followed by a specially designed lamella settler to collect resin. The collected resin is regenerated and reused. Resin is dosed from a dosing tank (concentrations of 4–20 mg/L depending on raw water conditions) into an open basin for fluidization of resin (air is often injected to assist in the kinetic transfer of ions) and under plug flow conditions, followed by settling in a special lamella settler. With the SIX process, all of the resin from the treatment basin/lamella settler is sent to regeneration and then is cycled back to the dosing tank for reuse in the treatment basin. Unlike the MIEX system, the SIX process can use almost any conventional anion exchange resin, and all resin that has been contacted with treated water is sent to the regeneration process before reuse in the dosing tank and then treatment basin. Figure 4-3 presents a schematic of the SIX process. The dose of regenerated resin to the treatment basin may be operator adjusted based on changing seasonal DOC conditions. Additional information about the SIX process, including a video, can be found on the PWNT website: https://pwntechnologies.com/portfolio-item/six-suspended-ion-exchange-process/.

4.4.3.3 Design Guidelines

Recommended design guidelines for MIEX:

Pretreatment:

- Non–scale-forming as indicated by LSI
- Oil and grease—not recommended
- Oxidants—most oxidants are tolerated within limitations
- No air bubbles; gases such as methane should be relieved before exposure to treatment process

System Design Guidance

Client Variables

- Treatment goal (in DOC, color, UVA reduction)
- Maximum flow/average daily flow/minimum flow
- Regenerate solution: sodium chloride

Manufacturer's Design Inputs

- Designated resin selected
- Resin concentration and CT in the fluidized bed
- BVTR (maximum under worse case and average) of specified resin
- Linear velocity for that specific resin (maximum hydraulic loading rate)
- Loading rate for plate settlers for that resin
- Salt dose per million gallons (MG) treated at 1,000 BVTR for that resin
- Waste brine volume per MG treated at 1,000 BVTR for that resin

Flow rates for municipal water treatment plants are generally well defined; therefore, treatment basin design for fluidized treatment systems may be based on maximum permitted rated flow of the plant and as dictated by state/provincial requirements regarding redundancy in pretreatment processes.

MIEX high rate process summary. The high rate process simply combines the resin fluidization and recovery steps into the same basin by stacking them vertically and using a higher concentration of resin to allow for higher velocities through the treatment process.

- MIEX resin contact time in fluidized bed: 3–6 min
- MIEX resin concentration in fluidized bed: 200 ml/L
- MIEX resin regeneration: sidestream regeneration skid provides consistent dosing/return of regenerated resin based on BVTR as determined by jar testing
- Linear velocity: 5.4–8.0 gpm/ft^2

PERFORMANCE NOTE

The maximum linear velocity of the process is determined by the resin manufacturer based on the properties of the resin selected and determines the carry over rate of the resin (resin lost to downstream processes). As the linear velocity approaches the minimum required to keep the resin suspended, less resin will be lost to carryover. In the high rate process using MIEX DOC resin, a linear velocity of 8 gpm/ft^2 yields an average resin loss rate of 1.2 gal/MG treated. This maintenance dose of resin prevents the process from requiring replacement of the entire inventory (as would be required in the fixed bed process at some interval of years) but is an ongoing operational expense that is determined during the design of the process.

Figure 4-4 shows influent lateral distributors, mixers, plate settlers, air lift pumps, loaded resin tank.

Figure 4-5 is an illustration of a MIEX high rate process inserted as pre-treatment for an existing coagulation/ballasted flocculation/settling basin. The drawing illustrates the high rate treatment basin ("resin contactor") airlift pumps for moving resin from the treatment basin to a loaded resin tank for dewatering before flow by gravity into the sidestream regeneration tanks for regeneration/pumping back to the treatment basin.

Movement of resin occurs via the following mechanisms:

- Airlift pumps transfer the resin from the treatment basin(s) to the loaded resin tank.
- Loaded (partially exhausted) resin will be dewatered in the loaded resin tank and then will flow by gravity from the loaded resin tank to the regeneration vessels for regeneration.
- Regenerated resin will be pumped from regeneration vessels back up to the top of the treatment basins using low shear pumps.

Airlift pumps use the buoyant force of air as the driving force for pumping resin slurry, which means there are no moving parts to maintain and lower resin attrition rates. Operation of the airlift pumps is controlled automatically and can be set to run either intermittently or continuously.

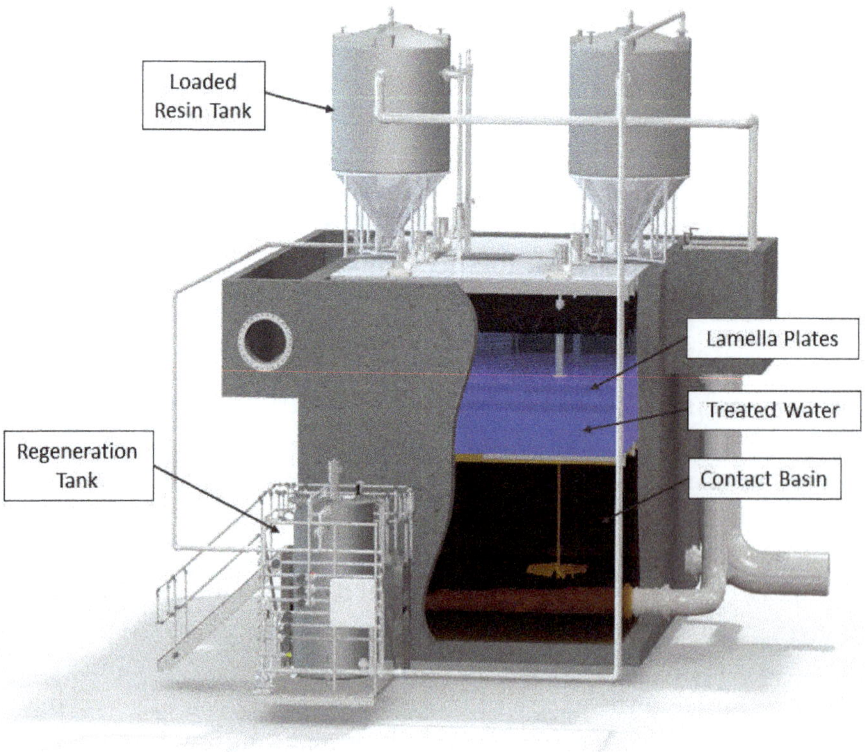

Figure 4-4 Illustration of a high rate treatment basin

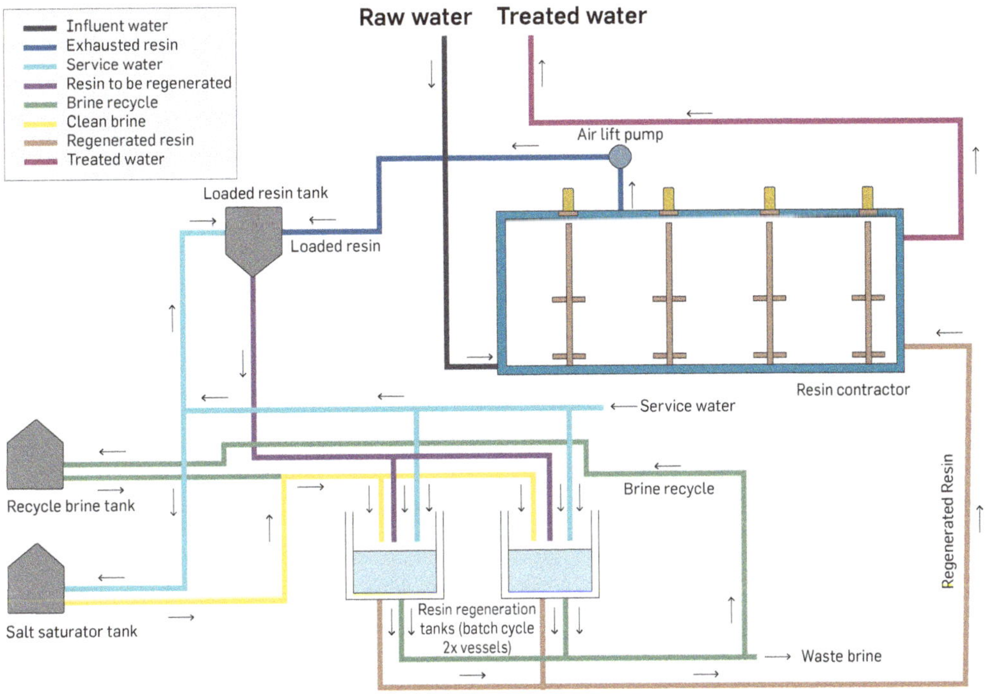

Figure 4-5 MIEX high rate process schematic

A key component of the above designs is the sizing of the sidestream regeneration process for regeneration of the active dose of resin required to maintain consistent DOC reduction. Due to the differences in composition of the DOC in each water, the removal efficiency and active dose of resin should be established using a bench test to determine if the fluidized bed system can achieve the minimum amount of DOC removal and at what active dose. Upon confirmation that the fluidized bed system achieved the desired DOC removal, the system may be scaled up for budgetary purposes and establishment of preliminary capital and operational costs.

Oftentimes, a pilot study may be performed to confirm the results of the bench testing and to assess the performance based on diurnal and seasonal variation of the raw water to optimize the design parameters and to quantify the characteristics of the brine waste. The jar test procedure is used to determine the optimum dose of resin required to sustain treatment goals. The treatment basin sizing is determined by the delivery model (High Rate, Dual Stage, Integra) and the hydraulic profile of the plant. The Multiple Load BVTR jar test is the most conservative estimation of treatment under the defined raw water conditions; as in all the above delivery models, there is additional capacity of resin in the treatment basin in addition to the active dose returned for stable treatment goals.

4.5 Pilot Testing

Chapter 13 of this handbook discusses the general process of piloting an ion exchange system. This section focuses on the specific considerations for TOC removal. As with all bench and pilot testing, the test should mimic and be scalable to the anticipated full-scale system. Bench-scale testing can help determine the optimal empty bed contact time (EBCT) to conduct subsequent experiments, screen resins to identify best performing options, and demonstrate treatment performance through multiple loading and regeneration cycles.

A comprehensive water quality analysis should be performed that includes all of the anions and cations. Chapter 2 presents a table describing the typical water quality analyses. With TOC, both total and dissolved organic carbon should be analyzed. Furthermore, uranium, sulfate, pH, alkalinity, and nitrate all will affect the selection of the most appropriate anion exchange resin, and sampling should be performed to enable an assessment of the variability of these parameters.

Based on manufacturer modeling, a pilot study may be beneficial to provide full-scale design parameters, including surface loading rate, throughput, EBCT and waste characterization. The pilot column tests generally have multiple trains that allow the comparison of different resin types.

4.6 Design Example

The following presents an example for a fluidized bed system for a surface water system.

4.6.1 Fluidized Bed Example Project Background and Design Criteria

In this example, the raw water is an upland reservoir with DOC ranging from 4–30 mg/L. The raw water alkalinity is less than 20 mg/L as calcium carbonate. The plant's existing treatment uses coagulation in the form of ferric sulfate, super pulsators for clarification, followed by RGFs (rapid gravity filters) and manganese contactors. Sodium hypochlorite is used for primary and secondary disinfection. The plant has a maximum rated capacity of 15.9 mgd (60 mld) but could not meet water quality goals operating much above 10.9 mgd (40 mld). It was having difficulty achieving post chlorination required DBP concentrations seasonally.

The utility invested in a fluidized bed pretreatment system that allowed it to meet the treated water DOC goals (thereby meeting DBP regulatory limits) with a secondary benefit of reducing the solids loading rate on the filters (a consequence of lowering the coagulant dose and creating a more stable floc). This returned the plant throughput to the rated 15.9 mgd (60 mld) capacity during all seasons. Another secondary benefit was a reduction in ferric sulfate consumption of between 31 and 52 percent, which reduced residuals by an equal percent.

In this example, the client utilizes UVA254 (expressed in m^{-1}) as a surrogate for DOC as it is easy to measure. The utility established 30 percent reduction of UVA254 as a minimum goal for the fluidized bed pretreatment process. Table 4-3 summarizes the raw water quality and Table 4-4 summarizes the design parameters of the fluidized bed system.

Table 4-3 Fluidized bed raw water parameters

Parameter	Units	Raw Water Concentration
UVA (254)	m^{-1}	45–157
True color	CU	86–296
Manganese	mg/L	0.033–0.818
Iron	mg/L	0.624–3.050
Turbidity	NTU	1.4–19.4
pH		5.0–7.2
Alkalinity	mg/L as $CaCO_3$	<20
Temperature	Degrees C	2–17

UVA: ultraviolet absorbance, CU: color units, NTU: nephelometric turbidity units

Table 4-4 Fluidized bed main design parameters

Design Variable	Units	Value
Maximum daily flow	mgd	15.85
Average daily flow	mgd	12.4
Minimum daily flow	mgd	4.0
UVA reduction goal	%	≥30%
Resin Type—SBA MIEX gold		
Treatment basin linear velocity (resin specific) max	gpm/ft²	5.4
Number of treatment basins	quantity	2, each at ½ flow
Maximum BVTR	BV	650
Average BVTR	BV	800
Fluidized resin contact time required in basin	min	6
Treatment basin resin concentration in fluidized bed (avg)	mL/L	200 (or 20% v/v)
Fluidized resin bed height to achieve contact time (contact zone) minimum	ft	6.5
Loading rate of lamella plate required to further reduce any resin carry over. Length of plates 7–9 feet (supplier dependent)	gpm/ft²	0.5
Salt dose/MG treated for regeneration (NaCl) based on 1,000 BVTR and standard regen parameters	lb/MG	400
Waste brine per MG treated from each regeneration cycle (based on 1,000 BVTR)	gal/MG	450

BV: bed volume, BVTR: bed volume treatment rate, SBA: strong base anion

4.6.2 Bed Volume Treatment Rate

The jar testing (multiple load jar test) for this example determines the maximum bed volume treatment rate under worse case conditions (expressed as the volume of water that can treat a volume of resin in order to produce the desired treatment). It is recommended to perform seasonal jar tests to create a history of treatment conditions. For this example, Table 4-5 shows the UVA254 reduction for the worst-case raw water conditions at different bed volumes. Table 4-6 summarizes the resin dose rate by BVTR.

While 1,000 BVTR (lowest dose rate) could meet the client performance goal of 30 percent reduction in UVA254 in this example, the client wanted to design around 800 BVTR and 650 BVTR in case raw water conditions deteriorate. In case of further deteriorating water quality conditions in the future, additional regeneration tanks could easily be added.

4.6.3 Resin Dose Rate

Dose rate of regenerated resin required to continue to produce the desired BVTR is calculated using the following simple formula:

Dose rate = 1,000 gal water/BVTR

Table 4-5 Multiple load jar testing summary of performance characteristics

BVs Bed Volumes	UVA (cm^{-1}) Raw	MIEX% Reduction
1,000	0.891	30.5%
800	0.891	33%
650	0.891	36.2%
450	0.891	40.9%
200	0.7891	51.7%

Table 4-6 Resin dose rate by BVTR

BVTR (water/resin)	Resin Dose Rate (gal resin/1,000 gal)
1,000	1.00
800	1.25
650	1.54
450	2.22
200	5.00

4.6.4 Contactor Treatment Basin

4.6.4.1 Sizing—Design Calcs

The contactor treatment basin is an upflow design, and the resin has been selected based on jar testing. The supplier recommended a specific small bead Type 1 strong base anion exchange resin (MIEX Gold), which has a maximum linear velocity (5.4 gpm/ft²).

From this recommended linear velocity, one can easily determine the basin area required. Assume there will be two (2) parallel basins each treating half the flow.

Basin area = flow/linear velocity
Convert from mgd to gpm: 15.9 mgd × 694.4 gpm/mgd = 11,041 gpm
Flow per basin at 11,041 gpm/2 basins = 5,521 gpm
Basin area at 5,521 gpm / 5.4 gpm/ft² = 1,022 ft² at approximately **32 × 32 ft**

Basin height is a function of the expanded bed contact zone desired for efficient adsorption kinetics, plus the net height of the inclined plate modules for secondary confinement of resin plus freeboard.

Volume of resin required to achieve CT at resin concentration

Based on the average concentration of 200 mL/L (or 20% v/v) (found in Table 4-4), the required fluidized contact volume will be:

Contact volume (minimum) = flow × required resin contact time (found in Table 4-4)

$$\frac{5{,}521\ \text{gal}}{\text{min}} \times 6.0\ \text{min} \times \frac{1\text{ft}^3}{7.48\ \text{gal}}\text{min} = 4{,}429\ \text{ft}^3$$

Contact height of expanded resin = volume / [length × width]

$$4{,}429\ \text{ft}^3\ /\ (32\ \text{ft} \times 32\ \text{ft}) = 4.4\ \text{ft}\ \ (20\ \text{percent of resin})$$

This is a theoretical contact height if the water was perfectly flowing upward in a uniform matter across the entire basin area—which is not the case. For large basins of this size, water is introduced via a series of influent distributor pipes and discharge nozzles which are directed downwards. Mixers are also used to supplement resin suspension at lower raw water flows—the combination of this energy will cause the resin to fluidize more than this minimum height needed. *Experience has shown that at a minimum, a design height of 6.50 ft of expanded resin height is required to fully fluidized the resin (@20%v/v)*[2].

$$32\ \text{ft} \times 32\ \text{ft} \times 6.5\ \text{ft} = 6{,}656\ \text{ft}^3 \times 7.48\frac{\text{gal}}{\text{ft}^3} = 49{,}787$$

20 percent of the basin volume is resin.

$$49{,}787 \text{ gal} \times 20\% = 9{,}957 \text{ gal of resin}$$

The top of the fluidized resin bed is not a dead flat layer. It will have a tendency to be a 'wavy' or ripple interface which is constantly moving and can vary ±6-9 inches in height. Between the maximum height of the expanded resin bed and the underside of the inclined plates, there is a need for a separation or quiescent zone, *which at a minimum would be ~2.0 ft.*

The height of basin due to lamella plates:

Most manufacturers use standard length plates varying between 7 and 9 ft in length. Figure 4-6 presents a sectional view of a plate settler. For this example, plate sizes of 3 ft-2 in. wide by 8 ft long are used. Inclined plates are typically installed at an angle of 55 degrees, which gives an overall vertical plate height of 6.55 ft (sin 55° × 8.0 ft). Taking into account the support frame thickness (~3 in.), and the 2–3 in. of water above the top of the plate and height of the baffles (~4 in.), *this gives an overall vertical height of ~7.5 ft for the incline plate module (will vary slightly based on plate manufactures design).*

Height of basin due to freeboard for weirs and overflow:

Most applications will want to allow 2.5–3.0 ft minimum freeboard from the top of the plates (water level) to the top of the basin wall.

Calculated minimum basin height:

Contact zone	6.5 ft (volume where resin is in contact with raw water)
Separation zone	2.0 ft
Settling zone	7.5 ft (net height of incline plate settlers and effluent collection)
Freeboard	3.0 ft (free board above water level to top of concrete)
Total height	19.0 ft

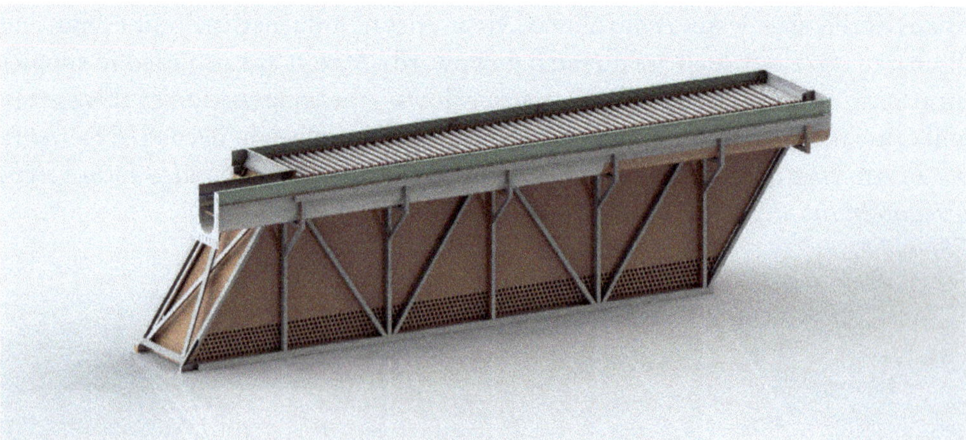

Figure 4-6 Lamella plate settler (section view)

NOTE: A significant factor that also influences the height of the basin is the overall plant hydraulics. As flow is gravity to downstream processes, this will have a decision in determining the overall height of the basin. In the "actual" basin heights below, the overall basin height was taller to account for some site specific hydraulics, and some additional conservatism.

Client's actual requested basin height:

Contact zone	9.0 ft (volume where resin is in contact with raw water)
Separation zone	3.0 ft
Settling zone	7.5 ft (net height of incline plate settlers and effluent collection)
Freeboard	3.5 ft (free board above water level to top of concrete)
Total height	23.0 ft

4.6.5 Resin Regenerated Volume

From testing, it was shown that the worst-case treatment rate of 650 bed volumes would achieve the desired treated water quality. This means 650 gal of water can be treated by 1 gal of resin. The daily volume of resin to be regenerated can be calculated as follows:

Resin volume = [15.9 mgd] × 1,000,000 mgd/gpd / [650 gal / 1 gal resin] = 24,462 gal resin regenerated per day

4.6.6 Regeneration Tank Sizes

Resin is taken from the main treatment contact basin and regenerated in smaller batches. The size and quantity of regeneration tanks is determined. Table 4-7 provides various diameter tanks along with the volume of resin each tank can regenerate per day (typically provided by equipment supplier), which determines the number of tanks that would be required on site in order to regenerate the required amount of resin (one must round up to the nearest whole tank). The fewest number of whole tanks on site that give the desired amount of redundancy, provides the lowest capital cost and smallest footprint. In this example, five (5) tanks each 7 ft diameter could be used, or three tanks 12 ft diameter; however, 3 tanks 10 ft diameter would be the most cost-effective choice.

Number of tanks = resin volume / resin tank volume
24,462 gal / 8,550 gal/tank = **2.86 tanks (so 3 tanks at 1,175 gal each)**

In this example, if the plant was operating at <40 mld (10.6 mgd), the maximum design treatment rate (650 BV) could be achieved with just two (2) regeneration tanks, so inherently there will be a redundant tank. Similarly, in periods of better raw water quality the treatment rate may be reduced—for example, 1,000 BV, which again would only require two (2) regeneration tanks. Understanding plant and system factors like this will weigh into the decision about whether to include redundant regeneration capacity.

4.6.7 Loaded Resin Tank

The loaded resin tank is sized to hold sufficient resin for two to three regeneration tanks full of resin (see Table 4-7). In this example there are three tanks, each 1,175 gal, and the loaded resin tanks are sized for 2.5 regeneration tanks full. In this case, 2.5 tanks of resin (2.5 × 1,175 gal) plus the slurry water contained in the 200 ml/L resin/water prior to dewatering will result in the following:

2.5 × 1,175 gal resin = 2,938 gal of resin

The tank is sized to include volume of carrier water. Carrier water needs to be 70 percent of resin slurry.

2,938 gal of resin × 1.7 = 4,994 gallons. Therefore, use 5,000 gal loaded resin tank minimum capacity

4.6.8 Salt Saturator Tank

The manufacturer has stated that 400 lb of salt per million gallons treated will be required at 1000 BVTR (1 gal resin regenerated per 1000 gal of water treated). The maximum BVTR for this example is 650 BVTR. From Table 4-6, Resin dose rate by BVTR, for 650 BVTR, the resin dose rate is 1.54 gal resin regenerated per 1,000 gal treated.

Salt for regeneration = resin dose rate × salt/MG treated

1.54 gal of resin/1,000 gal treated × 400 lb of salt/MG treated = 616 lb of salt / MG of water treated

15.9 MG / day × 616 lb/MG = 9,794 lb of salt per day max case = 4.9 ton salt/day

Salt saturators normally sized to hold 30 days of salt are equal to 150 tons of salt storage. Two 80-ton salt saturators (160 ton total capacity) would provide redundancy.

4.6.9 Brine Waste

The manufacturer has stated that each MG of water treated will produce 450 gal of residuals for disposal/1 mgd of treated water at 1000 BVTR. This concentrated

Table 4-7 Regeneration tank sizes

Tank Diameter (ft) (all 10 ft high)	Tank Volume of Resin (gal)	Daily Capacity Resin Regen (gal/tank)	No. of Regeneration Tanks Required
7.0	576	4,930	4.96
8.0	752	6,220	3.93
9.0	952	7,370	3.32
10.0	1,175	8,550	2.86
11.0	1,422	9,750	2.51
12.0	1,692	10,690	2.29

brine and organic mixture is often stored in a receiving tank and metered into a wastewater treatment plant for disposal. In this case, there was no sewer connection, so the liquid residual is hauled to a WWTP.

15.9 mgd × 450 gal of residual/MG treated × 1000/650 BVTR = 11,007 gpd = 330,210 gal/month

4.6.10 Resin Attrition

Some resin is lost from the contact basin and virgin resin will need to be dosed to maintain the desired resin concentration in the contact basin. The equipment supplier provided typical attrition rates based on flow rate.

Resin consumption per use:

- at average flow of 10.6 mgd, 0.3–0.5 gal of resin lost/MG of treated water
- at max flows of 15.6 mgd, 0.7–0.8 gal of resin lost/MG of treated water

4.6.11 Equipment Sizing Summary

Table 4-8 displays the process component sizing summary for the IX equipment. Total volume of resin in the system required at start-up:

Resin in contactor basins:

32 × 32 × 6.5 ft (contact zone depth) = 6,656 ft^3 × 20% of the volume is resin = 1,331 ft^3 = 9,957 gal × 2 contactor basins = 19,914 gal

Resin in regeneration tanks:

$\pi 5^2$ (tank area) = 78.5 ft^2 × 2.0 ft (resin bed depth) = 157 ft^3 = 1,175 gal/tank × 3 tanks = 3,523 gal

Resin on loaded resin tank:

5,000 gal × 1 tank = 5,000 gal

Total resin = 19,914 gal + 3,523 gal + 5,000 gal = 28,437 gal

Table 4-8 Process component sizing summary

Equipment Description	Material	Size (Approximate)	No. of Units
High rate contactors (HRC)	Concrete	32 × 32 × 19 (min) High	2
Regeneration tanks	FRP*	10 ft ID × 10 ft *h*	3
Loaded resin tank	FRP*	5,000 gal	1
Salt saturators	FRP*	80 ton	2

* Fiberglass Reinforced Polymer

Benefits to the WTP of this fluidized anion exchange pretreatment:

- Greater than 30 percent reduction of UVA (on average 40 percent)
- Reduction in ferric sulfate coagulant of 31–52 percent
- Reduction in solids loading rate commensurate with reduction in ferric sulfate
- Increased water production from the WTP during high DOC seasons of up to 33 percent
- Decrease in chlorine consumption of 20–30 percent approx. for same residual
- TTHM in final treated water <10 μg/L (record low)

See the Case Study in the back of the handbook for pictures and reference information on this example.

REFERENCES

Boyer, T.H. 2015. "Removal of Dissolved Organic Matter by Magnetic Ion Exchange Resin." *Current Pollution Reports* 1 (3): 142–54. https://doi.org/10.1007/s40726-015-0012-2.

Edzwald, J.K. 2011. *Water Quality and Treatment: A Handbook on Drinking Water, Sixth Edition*. 6th ed. AWWA, American Society of Civil Engineers, McGraw-Hill. https://www.awwa.org/Store/{ProductName}/ProductDetail/{ProductId}?productId=6292.

Erlandsson, M., Cory, N., Fölster, J., Köhler, S., Laudon, H., Weyhenmeyer, G.A., and Bishop, K. 2011. "Increasing Dissolved Organic Carbon Redefines the Extent of Surface Water Acidification and Helps Resolve a Classic Controversy." *BioScience* 61 (8): 614–18. https://doi.org/10.1525/bio.2011.61.8.7.

Galletti, C.J. 2000. "Just the Facts: Knowing Strong Base Anion Resin Types." Water & Wastes Digest. December 2000. https://www.wwdmag.com/just-facts-knowing-strong-base-anion-resin-types.

"Organic Compounds Definition, Examples, List, and Structure." n.d. *Chemistry Dictionary*. Accessed August 10, 2020. https://chemdictionary.org/organic-compounds/.

USEPA (US Environmental Protection Agency). 2007. "Drinking Water Treatability Database." Drinking Water Treatability Database. March 2007. https://tdb.epa.gov/tdb/contaminant?id=1175876466.

CHAPTER 5
Radionuclide Removal

5.1 Introduction

Radionuclides are radioactive forms of elements. They can be man-made or naturally occurring in the environment. In the context of water treatment, the most prevalent form of radionuclides are radium and uranium naturally occurring in groundwater. Generally, radium and uranium concentrations in source water are very low, but some localities can have levels that need treatment based on their geology. According to EPA:

> *Low-level exposure to radionuclide that may be found in water sources do not cause immediate health effects, but prolonged exposure increases risk of cancers, possible kidney damage and potential birth defects.*
>
> (Source: USEPA)

This chapter describes how ion exchange and single-use media systems can be used for the removal of naturally occurring radium and uranium in groundwater. This chapter describes:

- properties and chemistry associated with radium and uranium,
- regulatory issues associated with radium and uranium,
- source water quality parameters that can impact the selection of ion exchange resins suitable for radium and uranium removal,
- an overview of bench and pilot testing requirements,
- design guidelines including calculations and residuals generation, and
- an overview of the safety issues associated with the accumulation of radium and uranium on the resin.

5.2 Radionuclide Chemistry

In most naturally occurring groundwater applications, radium is found as a *cation*, and uranium is found as an *anion*.

Radium has an atomic number of 88 and exists as a divalent cation with similar characteristics as that of barium. The main two isotopes found in the environment are radium-226 and radium-228 with atomic weights of 226 and 228, respectively. Barium is a highly selective cation, second only to radium; as a result, the barium concentration of the raw water has the largest impact on the operating capacity of the strong acid cation (SAC) radium removal resins. Other cations, such as calcium or magnesium, are much less selective and as such have little to no impact on the radium capacity.

The selectivity strength of naturally occurring cations in groundwater follows the general relationship indicated here, with the lesser selective cation to the left and the higher selective cation to the right.

$$H < Na < NH_4 < K < Mg < Ca < Mn < Pb < Ba < Ra$$

In typical groundwater applications, a pH range of 4.0–8.0 will not adversely affect this ion exchange selectivity hierarchy.

Uranium has an atomic number of 92 and usually exists as an anionic uranyl carbonate, or uranyl sulfate complexes that have either a divalent or tetravalent ionic charge. This form of uranium has a strong affinity for strong base anion (SBA) resins. Operating capacity for uranium of the SBA resin is most heavily impacted by the presence and concentration of sulfate in the water because SBA resins also have a high selectivity for sulfates. Once the sulfate level exceeds the range of 50–75 mg/L, an additional evaluation must be done, including an evaluation of TDS and alkalinity.

The relative order of affinity of SBA resins for some common ions in drinking water show uranium at the top of the list with sulfate and chromium following:

Uranium/Perchlorate >> Sulfate/Chromium > Selenium/Arsenate > Nitrate > Chloride > Bicarbonate > Fluoride

When using a synthetic cation or anion resin in a single use configuration, there is potential for iron fouling. Therefore, the raw water should be analyzed for both soluble and insoluble iron. Soluble iron has the potential to foul the individual resin beads and insoluble iron has the potential to foul the vessel bed with suspended solids.

5.3 Radionuclide Regulations

5.3.1 USEPA Regulations

The original Radionuclide Rule was proposed in July 1991, but court action delayed its final promulgation. The final Radionuclide Rule was published in the Federal Register on December 7, 2000 and became effective on December 8, 2003.

In the final rule, USEPA set the maximum contaminant level (MCL) for uranium at 30 micrograms per liter (µg/L), using its authority under the Safe Drinking Water Act (SDWA) for the first time to set a standard at a higher than feasible level based on cost-benefit considerations. The standard for combined radium-226/228 remains at 5 picocuries per liter (pCi/L). However, the rule requires improved monitoring for radium. The final rule retains the interim standards for gross alpha particles at 15 pCi/L and for beta and photon emitters at 4 millirems (mrem).

A summary of the final Radionuclide Rule follows. Table 5-1 also lists the existing (1979) and the revised MCLs of the final Radionuclide Rule.

Affected systems: community water systems (CWSs); non-CWSs, including transient and non-transient, are exempt.

Maximum Contaminant Level Goals (MCLGs) for radionuclides: MCLGs of zero, includes:

- Combined radium-226/228
- Gross alpha
- Beta particle and photon radioactivity
- Uranium

Radium MCL:

- Combined Ra-226 and Ra-228 MCL of 5 pCi/L
- Based on new risk levels

Table 5-1 Previous and revised MCLs for radionuclides

Contaminant	1979 MCLs	2000 Radionuclide Rule MCLs
Radium 226/228	5 pCi/L	5 pCi/L
Uranium	N/A	30 µg/L (20.1 pCi/L)
Gross alpha	15 pCi/L	15 pCi/L
Beta particles and photon emitters	4 mrem	4 mrem

Beta/Photon Radioactivity MCL:

- ≤4 mrem/y to the total body or any given internal organ except for H-3 and Sr-90.
- H-3 = 20,000 pCi/L
- Sr-90 = 8 pCi/L
- Total dose from co-occurring beta/photon emitters must be ≤4 mrem/y to the total body of any internal organ.

Gross Alpha MCL:

- 15 pCi/L excluding uranium and radon but including Ra-226
- Maintain current MCL

Uranium MCL:

- 30 ug/L; new MCL

Proposed Radon Rule

Radon is a naturally occurring, carcinogenic, radioactive gas. Radon in drinking water increases risk to public health, primarily from inhalation of radon discharged through normal household use, such as showering, but also from ingestion of water. The proposed Radon Rule applies to all CWSs that use groundwater or mixed groundwater and surface water supply sources.

On November 2, 1999, the long anticipated and heavily debated Radon Rule was formally proposed, but USEPA missed the SWDA deadline of August 2000 promulgation. The USEPA is currently reviewing the US Government Accountability Office (GAO) report on the cost of the proposed rules.

The proposed rule includes a two-option approach that allows states and water suppliers to reduce radon risks in indoor air while protecting public health from the highest levels of radon in drinking water.

The proposed rule includes the following provisions:

■ Maximum contaminant level goal (MCLG)	zero
■ Maximum contaminant level (MCL)	300 pCi/L
■ Alternative maximum contaminant level (AMCL)	4,000 pCi/L

The AMCL provision of the rule applies to water systems that adopt and comply with a multimedia mitigation (MMM) program aimed at reducing household indoor/air health risks from the soil as well as the tap water. The AMCL of 4,000 pCi/L is based on the National Research Council recommended estimate of 10,000 to 1 as the transfer factor from water to air and the national average outdoor radon concentration of 0.4 pCi/L in air. Thus, an estimate of 0.4 pCi/L in air would be equivalent to 4,000 pCi/L in water.

If a state develops an MMM program that is approved by the USEPA, public water systems in that state will be able to comply with the AMCL rather than the MCL. Alternatively, if a state chooses not to adopt its own MMM program or a state's MMM program does not meet USEPA approval, an individual public water supplier can submit an MMM program for approval. The 1996 SDWA Amendments require that the USEPA evaluate MMM programs every five years. Although there is no enforceable standard for radon, radon can be found with radium and uranium. If treatment is being considered for radium and/or uranium, consideration should be given to the impacts of radon on the treatment system.

5.3.2 World Health Organization Guidelines

The World Health Organization (WHO) does not have specific regulations but rather guidelines based on gross alpha and beta activity. WHO has established screening values for alpha and beta activity of 0.5 Bq/L for gross alpha activity and 1 Bq/L for gross beta activity. Becquerel or Bq is the unit of radioactivity in the SI unit system, corresponding to one radioactive disintegration per second. Water should be screened for both gross alpha and beta activity. If the measured activity concentration is below the screened level, no further action or treatment is needed. If the concentration for either gross alpha or beta are above the screening level, the concentrations of individual radionucleotides should be determined and compared to the WHO guidelines for common natural and artificial radionuclides. These guidelines represent the concentration that if consumed through the year, would result in an individual effective dose of 0.1 mSv. Sieverts, or Sv, is the SI unit for effective dose, which accounts for the differences among different types of radiation so that biological impact can be compared directly.

5.4 Ion Exchange Resin Treatment Options

As with all water treatment applications, it is preferred to have a detailed raw water analysis before media selection and sizing. For the selection of radionuclide removal resins, the water analysis should contain the following constituents: combined radium 226 and 228, uranium, gross alpha, barium, sulfate, iron, manganese, pH, TDS, TSS, alkalinity, radon, and water temperature.

5.4.1 Radium 226 and 228

As described previously, in groundwater applications, radium occurs as a divalent cation and is one of the most highly selective cations for resins. Therefore, SAC resin in the sodium (Na^+) form will satisfactorily remove radium from relatively high background levels. Barium is also a highly selective cation, and raw water levels should be checked to determine ultimate radium removal capacity. Other raw water parameters that may impact radium removal include iron, hardness, and pH. Iron can cause fouling of the resin, and pretreatment for iron removal may be needed. Hardness can be a competing ion for the resin, and overall removal efficiency will be affected. Additionally, pH impacts reaction kinetics, and for optimal resin performance, pH should be between 5 and 8.5. Levels outside of this range should be actively monitored to make sure that acceptable performance of the resin is achieved.

Because there is no online or real-time radium analytical instrumentation, an ion exchange system is typically regenerated based on hardness breakthrough or on a precalculated throughput volume. Basing system regeneration on hardness breakthrough provides several beneficial operational features. It ensures the treated water radium level remains well below the MCL, hardness monitoring is relatively easy, and that the regenerant waste solution radioactivity level stays low.

By basing system regeneration on a calculated throughput, you maximize run time between regenerations, thereby lowering operating costs, but have a concentrated radium laden waste that is sent to the receiving sewer.

5.4.2 Uranium 234, 235, and 238

As described previously, in groundwater applications, uranium occurs as a multivalent anion and is one of the more selective anions for resins. Therefore, SBA resin in the chloride (Cl-) form will satisfactorily remove uranium from relatively high background levels of other anions. Raw water sulfates can hinder maximum uranium capacity, so levels should be checked to determine ultimate uranium removal capacity. Other raw water parameters that may impact uranium removal include alkalinity, iron, and pH.

There is no online or real-time uranium analytical instrumentation. Therefore, an ion exchange system is typically regenerated based on sulfate breakthrough or on a precalculated throughput volume. Basing system regeneration on sulfate breakthrough ensures the treated water uranium level remains well below the MCL and that the regenerant waste solution radioactivity level stays low.

By basing system regeneration on a calculated throughput, you maximize run time between regenerations, thereby lowering operating costs, but have a concentrated uranium laden waste that is sent to the receiving sewer.

5.4.3 Combined Radium and Uranium Removal

If a source water has both types of radionuclides, radium and uranium can be removed in the same reactor by passing through a mixed bed of both types of resin: cation to remove radium, and anion to remove uranium. Alternatively, two tanks, one of each type of resin, can be configured in series. This configuration offers the benefit of removing dissolved iron and manganese in the cation tank that could foul the anion resin. The same brine solution can be used to regenerate both cation and anion media. If using a single use resin or media, the two contaminants must be treated in separate vessels. This is needed because of licensed disposal considerations and the specific licensing of those disposal sites.

5.5 Bench and Pilot Testing

As with all bench and pilot testing, the test should mimic and be scalable to the anticipated full-scale system. Use as large a column as convenient so that the column wall forces are negligible, typically 4–6 in. in diameter. For regenerable pilot systems, onsite regeneration may not be possible due to the inaccessibility of a sewer connection.

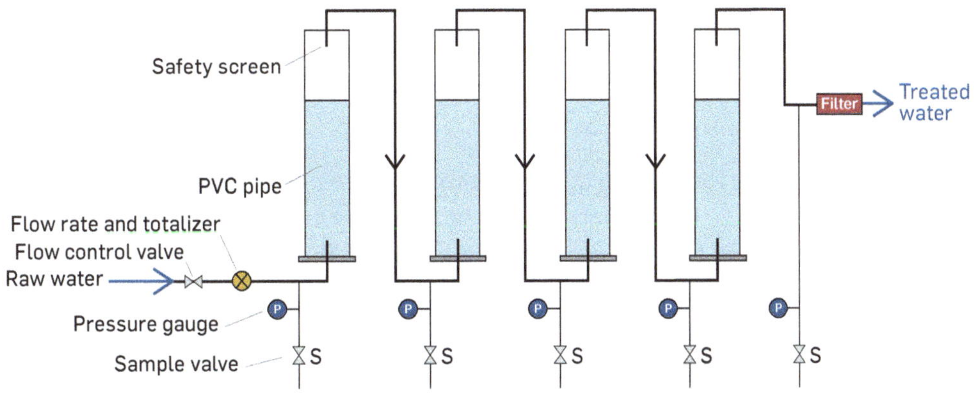

Figure 5-1 Pilot testing schematic for single-use media

Pilot testing of a single use media is easier than a regenerable media as the system is passive and doesn't require a sanitary sewer to receive the regenerant waste solution. When running a single-use media pilot test it is important to run at least one column to exhaustion so that media throughput and then the cost of long term operation can be calculated. Figure 5-1 presents a pilot system schematic.

When performing any radionuclide removal, check with the applicable state regulatory agency as a state specific radioactive materials license (RML) may be required to perform an onsite pilot test. Also, the test media used should be disposed of in a licensed manner.

5.6 Design Guidelines

Recommended design guidelines for regenerable resins are further described below. The brine regenerant concentration and dosage can vary depending on the raw water contaminant level. The higher the contaminant level, the higher the concentration and dosage.

The recommended design guidelines for Radium removal are:

- Resin: strong acid cation in the Na^+ form
- Empty bed contact time: 3.5–7 min
- Hydraulic loading rate: 4–8 gpm/ft^2
- Resin depth: 3–4 ft
- Regenerant solution: sodium chloride brine
- Regenerant concentration: 5–10 percent NaCl
- Regenerant dosage: 5–10 lb/ft^3

The recommended design guidelines for uranium removal are:

- Resin: SBA in the Cl^- form
- Empty bed contact time: 3.5–7 min
- Hydraulic loading rate: 4–8 gpm/ft^2

- Resin depth: 3–4 ft
- Regenerant solution: sodium chloride brine
- Regenerant concentration: 5–10 percent NaCl
- Regenerant dosage: 5–10 lb/ft^3

For single-use resin, the system design parameters are not as easily determined or generalized as those for a regenerable ion exchange system. The sizing and design consists of optimizing the volume of media to reduce the frequency of changeout, thereby reducing the long-term operating costs while minimizing the vessel size to reduce capital cost.

The volume of media is determined based on the raw water radionuclide and competing contaminant concentrations, the well pump flow rate, selection of the appropriate empty bed contact time (EBCT), and a practical or desired media exchange frequency. The EBCT varies depending on the raw water contaminant level and can range from 4.5–9 min. Once the volume has been determined, the number of vessels can be determined. Surface loading or hydraulic loading rate (HLR) is an important parameter in selecting the vessel size and is generally in the 5–8 gpm/ft^2. Too high a rate will result in a high pressure drop, and too low a rate will result in uneven flow distribution.

5.6.1 Pretreatment Requirements

Pretreatment requirements depend on the specific media or resin selected for contaminant removal, and those requirements vary depending on the specific media chosen. Before settling on one specific media, discussions with the media supplier should be undertaken, or in the case of limited data, pilot testing may be needed.

If a synthetic resin is chosen, there can be resin fouling with raw water iron levels as low as 0.2 mg/L. In that case, iron sequestration prior to radium removal by ion exchange can mitigate or at least slow fouling. If raw water iron levels exceed that of sequestration (~0.3 mg/L), additional treatment may be needed. However, it should be noted that radium will typically co-precipitate with iron and/or manganese, so any discharge or backwashing of an iron and manganese filtration unit can create a radium-laden waste. Depending on the applicable state or federal wastewater discharge regulations, that discharge may not be allowed.

If a zeolite-based radium removal media is used, no pretreatment for iron or manganese is needed as the media is transparent to those raw water constituents, and the presence of iron or manganese will not impact the ultimate radium capacity.

Uranium removal resins, being SBA resins, are especially susceptible to fouling from iron, manganese, and organics. An analysis of the overall raw water and discussion with the media supplier is needed to determine the impact on ultimate uranium capacity.

5.7 Disposal Options

The radionuclide residuals from any treatment technology are designated as technologically enhanced naturally occurring radioactive materials, or TENORM.

Depending on the specific state or federal regulations, sewer disposal of TENORM treatment residuals may or may not be an option. In states where sewer disposal was initially allowed, the regulations are evolving as more is learned about long-term wastewater treatment plant operation and any unintended consequences of radionuclide residual disposal, so state regulators should be contacted regarding the applicability. And even in cases where a sewer disposal is allowed under state regulations, the receiving sewer authority has the option to reject receipt of that regenerant waste. If sewer disposal is contemplated, discussion and written acceptance with the sewer authority should be received before system design.

If sewer disposal is an option, the ion exchange regenerant waste should be in solution with no discrete radioactive particles. Typically, a liquid waste with discrete, hard particulate is not allowed to be discharged to the sewer. The discrete particles would exclude the non-radiological suspended solids generated during the initial backwash step of regeneration.

For single-use resin systems, there will be a point when the media is exhausted and will need to be replaced. That exhausted media is now a low level radioactive waste (LLRW) and must be disposed of in a facility licensed and permitted to handle that specific waste. The permitted disposal site will hold a license that will allow acceptance of waste up to certain radium activities. There are three levels of combined radium activity depending on the disposal facility license: (1) <200 pCi radium activity / gram of media; (2) <1,500 pCi/g; and (3) <9,999 pCi/g

With each level of activity, there are specific packaging and handling requirements as well as different cost structures. Figure 5-2 describes the basic waste disposal options.

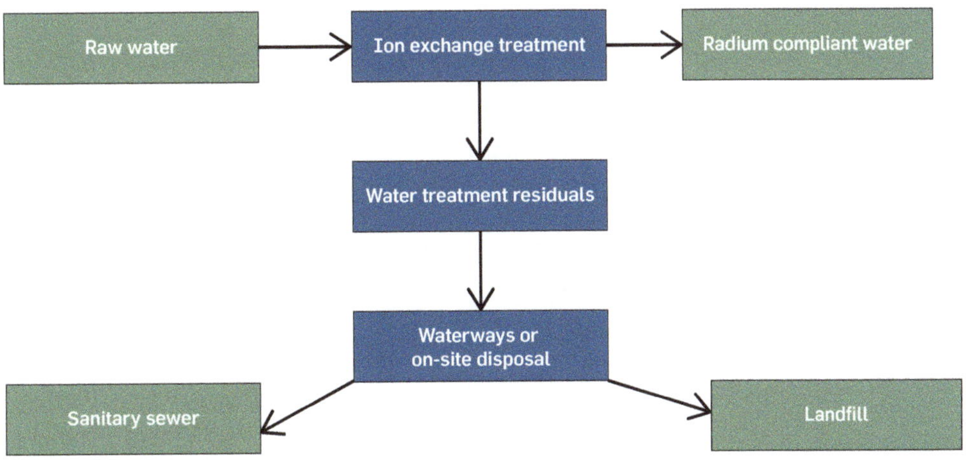

Figure 5-2 Waste disposal options

The disposal facilities will require a physical media sample analysis for radioactivity or demonstration of the media activity before it is accepted for disposal.

Onsite storage of spent media before release and shipment for disposal can also vary state by state. The amount and time allowance will be determined by the specific state RML and may be detailed as one of the RML conditions.

Uranium can be disposed of in the same manner as radium, although there are different disposal facilities, or different locations within the same facility, where the uranium-laden media can be managed. Unlike radium, uranium does have a beneficial use, so it can be recycled. Once removed from the treatment system, the uranium-laden media is shipped to a mill where the uranium is stripped from the media. The stripped media can then be disposed of in a local landfill or under certain circumstances it can be reprocessed for future use at an NSF-certified facility.

Both radium and uranium treatment residuals fall under the Department of Transportation (DOT) regulations. When either is being shipped to a disposal site or moving between utility locations, a DOT transportation specialist should be consulted as the shipment will be a DOT Class 7 haul, which will require a designated truck regardless of the spent media volume.

As would be expected, transportation and disposal of treatment residuals is a significant cost of long-term system operation.

5.8 Radioactive Materials Licensing

Depending on the state in which the system will operate, there may be a requirement to obtain and maintain a specific radioactive materials license (RML), regardless of the removal technology chosen.

For radium removal systems, there is almost always a state-specific RML required regardless of the treatment system or utility size. Licensing differs from state to state, so it is best to check with the state regulatory agency to determine the necessity of licensing and the type required. If a specific RML is required, it must be in place before treatment system commissioning. To date, only two states do not currently require a specific RML for radium removal system operation: Missouri and North Carolina.

USEPA provides the regulation basis, but an individual state may also choose more strict regulations.

Considerations for treatment system size and subsequent licensing include:

- Raw water radium level
- Treatment system size and media volume
  - Treatment system changes, increases in regeneration frequency, or water usage changes
- Anticipated radium inventory
  - For regenerable systems, the inventory just prior to regeneration
  - For single use systems, the inventory just prior to a media exchange

- Radiation protection program
  - Radiation monitoring of operations personnel
  - Radiation awareness training of operations personnel
  - State required postings at the treatment facility
  - Annual third-party program audit
- System radiation monitoring, typically in five locations
  - A monitoring badge on each treatment vessel
  - A near system area badge
  - An off-site control badge
  - A badge on the treatment site fence line or the closest location of access to the general population

Badging for both personnel and the treatment system typically consist of an optically stimulated luminescence (OSL) badge similar to the Luxel model supplied by Landauer (Glenwood, Ill.). The OSL badges are usually changed out on a quarterly basis.

With specific RML, there also may be a state requirement for financial assurance (FA). FA is a financial vehicle put in place between the utility and the state to ensure financing of the end of system decommissioning and decontamination if the original equipment and service provider is not able to provide that service.

For uranium removal systems, there are two types of RMLs: a general license or a specific license.

A small system (typically <40 gpm) can fall under the general licensing conditions if the on-site uranium inventory stays below 15 lb. For larger systems, a specific license is required. That specific license would carry the same conditions and restrictions as that for radium.

5.9 Design Example

A well with a capacity of 700 gpm contains both radium and uranium above the MCLs. A sewer is available on site. Table 5-2 presents the raw water characteristics. Because a sewer is available, the initial choice is a regenerable ion exchange system. If a sewer is not available or the receiving publicly owned treatment works (POTW) will not allow a radionuclide laden waste, a single-use system would be chosen. Given that there are both radium and uranium in the groundwater, a minimum of four vessels (two in series for radium, and two in series for uranium) will be chosen.

In this application, a sewer connection is available, so the raw water iron and manganese levels are not as critical. Any, or at least most, of the iron or manganese captured by the resin will be removed during the regeneration procedure. If in this example a single-use resin was chosen, the raw water Fe/Mn should be sequestered so that resin fouling is minimized and the ultimate radium capacity is retained.

As noted previously, uranium removal is accomplished using an anion resin, which can be susceptible to fouling from organics. In this application, the organics loading is low, so unacceptable resin fouling is not anticipated.

Table 5-2 Raw groundwater parameters

Raw Water Parameter	Value
Well pump flow rate	700 gpm
Firm treatment capacity	700 gpm
Sewer availability	Yes
Radium-226 + radium-228	27 pCi/L
Raw gross alpha	37 pCi/L
Uranium	45 µg/L
Adjusted gross alpha	7 pCi/L
Iron	0.3 mg/L
Manganese	0.01 mg/L
COD	4.5 mg/L
TDS	250 mg/L
pH	5.1

COD: chemical oxygen demand, TDS: total dissolved solids

A raw water radium level of 27 pCi/L is on the high end, so a longer EBCT is needed.

Radium resin volume: A 7 min EBCT is targeted

$$\frac{700\ gal}{min} \times 7\ min = 4{,}900\ gal \times \frac{ft^3}{7.48\ gal} = 655\ ft^3\ of\ resin$$

Since there will be two vessels in series: 655 ft^3 / 2 = 328 ft^3 of resin per vessel.

Vessel sizing: Assume a resin depth of 4 ft → 328 ft^3/4 ft = 82 ft^2 vessel cross-sectional area. Chose standard vessel diameter of 10 ft-6 in at 86.5 ft^2 cross-sectional area. HLR check → 700 gpm / 86.5 ft^2 = 8.1 gpm/ft^2, which is on the high end of the HLR range but acceptable. Resin depth check: 328 ft^3 of resin / 86.5 ft^2 area = 3 ft-9 in, which is well over the minimum depth.

Radium resin volume (metric): 7 min EBCT x 2,650 L/min = 18,550 liters empty bed volume/1,000 L/m^3 = 18.55 m^3 of resin. Since there will be two vessels in series: 18.55 m^3 / 2 = 9.28 m^3 of resin per vessel.

Vessel sizing (metric): Assume a resin depth of 1.2 m → 9.28 m^3/1.2 m = 7.7 m^2 vessel cross-sectional area. Choose standard vessel diameter of 3.2 m at 8.04 m^2 cross-sectional area. Make HLR check → 2,650 LPM / 19.6 m^2 = 135 LPM/m^2, which is on the high end of the HLR range but acceptable. Make resin depth check: 9.28 m^3 of resin / 8.04 m^2 area = 1.2 m+, which is well over the minimum depth.

In this case, with radium being the higher contaminant level, it becomes the limiting factor and the uranium system will be conservatively designed using similar vessel sizing and resin volumes.

With the raw water radium level on the high side, a higher regenerant brine concentration and dosage should be used for design purposes. After startup, the operator then has the option of adjusting the concentration and dosage to minimize long-term operating costs.

With the raw water uranium level on the lower end, a lower brine concentration and dosage can be used for design. Again, field adjustment can be made after system commissioning.

To estimate regeneration frequency and therefore daily regenerant waste volumes, the design engineer must have discussions with the client to better understand how the client would like to operate the system.

CHAPTER 6
Arsenic Removal

6.1 Introduction

Arsenic found in drinking water can come from natural or manmade sources. Arsenic is a naturally occurring element that can be found in rocks, soils, water, air, and plants and animals. Industrial uses of arsenic include pigments in paints and dyes, ingredients in agriculture products such as insecticide and herbicides. It is used in some drugs and metals, but in industry today about 90 percent of arsenic is used for wood preservation (USEPA 2001).

Arsenic is more prevalent in groundwater than in surface water sources. Arsenic can be very geography specific, appearing in localized "hot spots." In general, the western United States has more arsenic than the rest of the country (USEPA 2001).

This chapter describes how regenerable ion exchange systems can be used for arsenic removal for drinking water. This chapter describes:

- properties and chemistry associated with the forms of arsenic found in groundwater and surface water sources;
- regulations regarding arsenic: specifically the US Environmental Protection Agency (USEPA), the World Health Organization, (WHO) UK Water Supply Directive, and the European Commission Directive;
- source water quality parameters that can impact the selection of ion exchange resins;
- an overview of the types of resins that can be used for arsenic removal;
- an overview of bench and pilot testing requirements;
- design guidelines, including calculations and residuals generation; and
- a design example.

6.2 Arsenic Occurrence and Chemistry

6.2.1 Occurrence and Uses

Arsenic (As) is an element that occurs in the earth's crust, and groundwater can solubilize and erode naturally occurring arsenic. Arsenic can also enter water supplies from man-made activities such as:

- semiconductor manufacturing,
- petroleum refining,
- manufacture and use of wood preservatives,
- manufacture and use of animal feed additives, and
- manufacture and use of pesticides and herbicides.

During the production of metals such as copper and lead, arsenic wastes are often a byproduct (Loebenstein 1994). More information on arsenic production can be found in "The Materials Flow of Arsenic in the United States" by J. Roger Loebenstein, published by the US Department of the Interior.

According to the US Department of Health and Human Services 14th Report on Carcinogens, November 3, 2016, arsenic concentrations in untreated surface and groundwaters are highest in the western United States and lowest in the southeastern and mid-Atlantic portions of the United States; further, 12 percent of the groundwater systems in the West have groundwater above 20 ppb (NTP 2016).

WHO has found arsenic in groundwater to be a worldwide issue with occurrence of arsenic in more than 50 countries, including Argentina, Bangladesh, Chile, China, Hungary, India, Mexico, Nepal, and Taiwan (Ahmad, Khan, and Haque 2018).

6.2.2 Chemistry

Arsenic in water is found in two valence (or oxidation) states: arsenite, As(+3) also referred to as As(III) and arsenate, As(+5) or As(V). As(+3) species exist as uncharged compounds in water most commonly as arsenious acid (H_3AsO_3). As(+5) species form anionic compounds in water. Common forms of As(+3) and As(+5) are shown below:

As (+3) - $H_3AsO_3^0$, $H_2AsO_3^{-1}$, $HAsO_3^{-2}$
As (+5) – $H_3AsO_4^0$, $H_2AsO_4^{-1}$-, $HAsO_4^{-2}$
(Sorg and Lytle 2019)

6.3 Arsenic Regulations

This section describes the drinking water regulations and the health effects associated with arsenic. An overview of the USEPA's arsenic rulemaking history is provided as the revisions to the arsenic rule can be valuable in understanding the impacts of arsenic on public health along with benefits of providing treatment.

6.3.1 USEPA Regulations and Standards

Consumption of drinking water with arsenic can cause both cancerous and non-cancerous health effects. The USEPA has classified arsenic as a known human carcinogen and can cause cancer of the bladder, lungs, skin, kidney, nasal passages, liver, and prostate. Non-cancer health impacts can include cardiovascular, pulmonary, immunological, and neurological effects (USEPA 2015b).

In 1975, USEPA established a drinking water standard of 50 ppb for arsenic. The 1985 Safe Drinking Water Act Amendments required USEPA to establish maximum contaminant levels (MCLs) and maximum contaminant level goals (MCLGs). In June 2000, USEPA proposed an MCLG of zero and an MCL of 5 ppb for arsenic. As part of the proposed rule, USEPA requested comment on establishing the MCL at 20, 10, and 3 ppb. In January 2001, USEPA published the final arsenic MCL of 10 ppb with the MCL being enforceable in January 2006. Due to concerns associated with the cost of the rule, in March 2001 USEPA authorized three studies to assess the health effects, costs, and benefits associated with the rule, and in October 2001 announced that the MCL would remain at 10 ppb and according to the USEPA administrator Christine Todd Whitman, the studies "have reinforced the basis for the decision" (USEPA 2015).

The MCLG is a non-enforceable public health goal and represents the maximum level of contaminant in drinking water at which no health effects would occur. Generally, all of the state regulators have adopted the USEPA arsenic MCL with the exception of the New Jersey Department of Environmental Protection (NJDEP), which established a more stringent level of 5 ppb MCL. As discussed previously, the NJDEP arsenic MCL is consistent with USEPA's originally proposed 5 ppb MCL.

6.3.2 International Regulations and Guidelines

WHO in its 1996 Guidelines for Drinking Water Quality, 2nd ed. provides a guideline of 0.01 mg/L (10 ppb) for arsenic (WHO 1993).

In Canada, the maximum acceptable concentration (MAC) for arsenic is 10 ppb in drinking water. According to Health Canada, "in the case of arsenic, the guideline is higher than the concentration that would present an "essentially negligible" risk of internal organ cancers, since it represents the lowest level of arsenic in drinking water that can be technically achieved at reasonable cost, especially for smaller public systems and private wells (Canada 2006).

WHO, United Kingdom Water Supply Regulations, and European Commission Directive 98/83 all have a 10 ppb standard for arsenic.

6.4 Ion Exchange Resin Treatment Options for Arsenic

As(+3) is generally poorly removed by most water treatment processes and oxidation to As(+5) is needed. As part of the arsenic rule, USEPA has identified best available technologies (BATs) for As(+5) removal, and the BATs are

- ion exchange,
- activated alumina,
- reverse osmosis,
- modified coagulation/filtration,
- modified lime softening,
- electrodialysis reversal, and
- oxidation/filtration.

(USEPA 2015a)

There was a lack of published data for additional adsorptive media beyond activated alumina at the time of the promulgation of the rule in 2001. However, over the past 20 years new single-use adsorptive media for arsenic removal have been developed. The application of single-use adsorptive media will generally depend on the exhaustion rate and cost. High sulfate levels compete vigorously for the same ion exchange site as arsenate. Ion exchange is generally suitable for removal of As(+5) under the following conditions:

- TDS is less than 500 mg/L.
- Sulfate is less than 150 mg/L.

(Edzwald 2011)

Ion exchange needs to be in the form of strong base anion (SBA) of Type 1 or Type 2 when using ion exchange for removal of arsenic. The removal of arsenic requires that the arsenic be in the As(+5) state and pre-oxidation of As(+3) to As(+5) is needed. Generally, chlorine is used for oxidation and for 1 ppb of arsenic +3, 3 ppb of chlorine is needed.

SBA exchange resins can remove more than 95 percent of arsenic. As with all water treatment applications, a detailed raw water analysis is needed for proper resin selection and system design.

With SBA resin, generally no pH adjustment is needed, and the resin can be reused multiple times by using a brine solution periodically to elute the arsenic from the resin. The SBA design must include proper treatment and disposal of the spent arsenic-containing brine.

Ion exchange is very effective at removing arsenic, but not 100 percent effective. The amount of arsenic that is not removed is called leakage and can vary based on the type of resin used and other constituents in the water. Typical leakage can be expected to be around 1–3 ppb. Since the leakage and MCL are in close proximity, blending is not recommended.

6.4.1 Strong Base Anion Exchange Resins

SBA exchange resins are effective over a wider pH range and do not typically require pH adjustment. Water passes through the resin where arsenic and other anions are exchanged for chloride initially.

The order of affinity of Type 1 and Type 2 SBA resins for anions typically found in water is as follows:

$$CrO_4^{2-} > SO_4^{2-} > HAsO_4^{2-} > NO_3^- > Cl^- > HCO_3^- > F^-$$

Depending on the water chemistry, chromium and sulfates will be removed more preferentially. A less preferred ion will be concentrated on the resin and can exit the column at a concentration higher than their influent concentration. This effect is known as chromatographic peaking.

A typical ion exchange reaction is

$$HAsO_4^{2-} + Resin - Cl_2 \rightarrow Resin - HAsO_4 + 2Cl^-$$

Regeneration typically is performed with sodium chloride, and the regeneration reaction is

$$Resin - HAsO_4^{2} + 2NaCl \rightarrow Resin\text{-}Cl_2 + Na_2\,HAsO_4$$

Resins available for arsenic removal are of two main types (Purolite 2015):

- Type 1 SBA resin
- Type 2 SBA resin

These resins are effective over wide ranges of temperatures and pH. Designers should refer to local drinking water approved product lists before selecting an ion exchange resin. Major differences in these types of resins are discussed in the next section.

6.4.2 Comparing Type 1, Type 2 Strong Base Resins

Type 1 SBA ion exchange resins are manufactured with trimethylamine functional groups while Type 2 resins are manufactured with dimethylethanolamine functional groups.

One drawback of Type 1 SBA resins is the tendency to give off a slight fishy amine smell that may be detected by some sensitive customers even at very low parts per trillion concentrations (Purolite 2015). The phenomenon is particularly noticeable in systems that experience frequent stop/start operation such as for small community water systems or point-of-entry (POE) household use. Therefore, Type 1 resins are generally not recommended for such service (Purolite 2015). Type 1 resins are, however, acceptable for use in municipal plants since these plants tend to operate continuously and the long distribution lines to consumers help to dissipate odor. Another good reason to consider Type 1 resins for municipal use is the availability in the last decade of higher capacity versions that tend to rival the economics of the other types of resins. More information on Type 1 resins is provided in Chapter 2: Fundamentals of Ion Exchange.

Type 2 SBA resins do not generally impart odor to the water and can be used across the entire spectrum of the municipal, small community, and POE systems.

For both Type 1 and Type 2 resins, the relative order of affinity for common anions is as follows:

$$SO_4^{2-} > HAsO_4^{2-} > NO_3^- > Cl^- > HCO_3^- > F^-$$

Type 1 and Type 2 SBA resins have a higher affinity for sulfate and arsenic than for nitrate. Passage of water through the ion exchange column results in sulfate being preferentially loaded at the inlet of the column with arsenic and other anions progressively pushed deeper into the ion exchange column. When sulfate reaches the end of the bed, it displaces the concentrated nitrate, which can exit the column at a concentration higher than the influent concentration. This is known as chromatographic peaking (or dumping) of nitrate and is discussed more extensively in Chapter 2. Treatment plants are therefore designed to stop the service cycle at a safe nitrate breakthrough point so peaking does not occur. At that point, it is important to use the correct amount of salt to regenerate the resin in preparation for the next service cycle.

An example of nitrate peaking before sulfate or arsenic (As[+5]) breaks through is shown in Figure 6-1 for a Type 2 SBA resin. In this example, nitrate is also present in the water and peaks before arsenic. Nitrate is less toxic than arsenic and can be used as a surrogate in determining when to end the run.

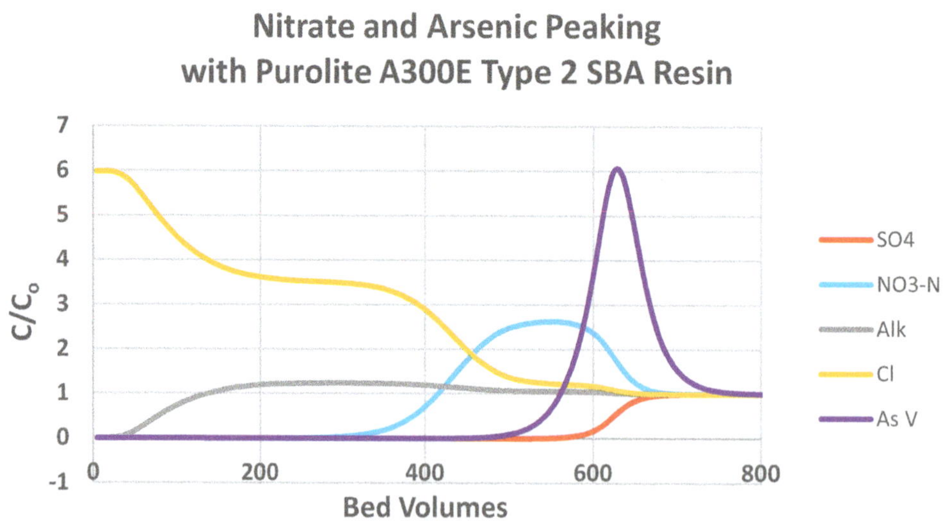

Figure 6-1 Nitrate peaks before arsenic with a Type 2 SBA resin

Courtesy of Purolite Corporation

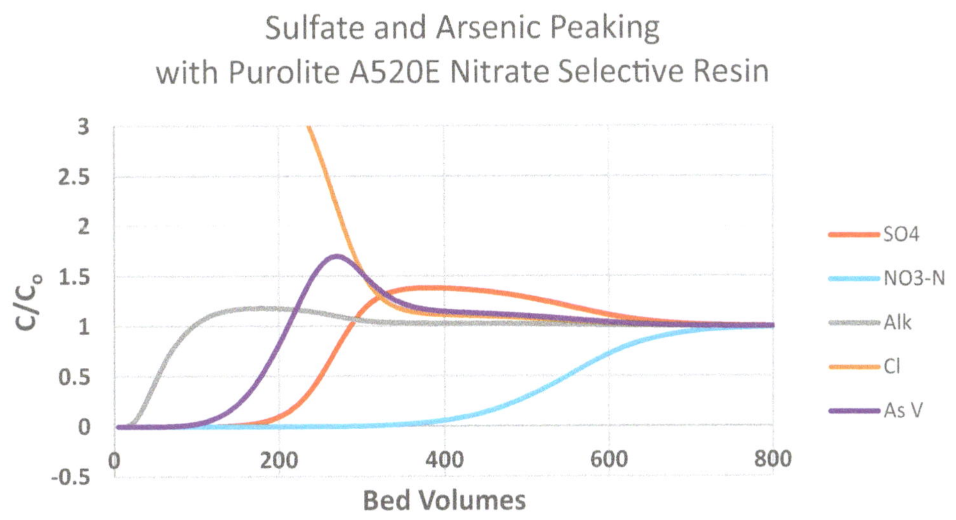

Figure 6-2 No nitrate dumping with a nitrate selective resin
Courtesy of Purolite Corporation

Determining the most cost-effective resin should consider the cost of the resin, the cost of regenerant per volume of water treated, and the cost of wastewater disposal.

In cases where simultaneous removal of arsenic and nitrate is needed, nitrate selective resins should not be used as arsenic will peak before nitrate breaks, as seen in Figure 6-2.

Since arsenic is more toxic than nitrate, it is preferable to use either a Type 1 or Type 2 SBA resin as nitrate breaks through before arsenic.

When a vessel of resin has reached its change-out criteria, the vessel is taken out of service in preparation for regeneration. During the regeneration step, several bed volumes (BVs) of sodium chloride (NaCl) brine regenerant are pumped through the resin at a high enough concentration to strip the arsenic and other anions from the resin and return the resin to chloride form. The brine will contain high concentrations of both sodium chloride and arsenic. The system then goes through a slow rinse to displace the last contaminants followed by a fast rinse, mainly to displace the NaCl. The vessel is then ready to be placed back in service.

6.4.2.1 Brine Disposal Options

In most cases, spent brine cannot be discharged to a sanitary sewer system due to the impact on the wastewater treatment plant. The impacts on wastewater treatment plant discharges and National Pollutant Discharge Elimination System (NPDES) requirements (USEPA 2007) must be evaluated for each system.

Brine reuse has been studied by determining where the arsenic concentrations are highest in the brine stream with an elution study (usually between 2 and 4 BV). The high arsenic portion of the brine waste can be disposed of and the rest of the spent brine can be reused. This practice reduces the volume of waste that needs to be hauled off. Disposal of the brine waste would generally consist of hauling the liquid waste to a hazardous waste facility.

High arsenic in the spent brine may result in it being classified as hazardous. Typically, a brine analysis is needed prior to disposal. As(+5) can be removed from the spent brine by precipitation using an aluminum- or iron-based coagulant. The settled sludge containing arsenic will be in the form of a liquid that can be sent to a sewage treatment plant or dewatered and with subsequent cake disposal (Edzwald 2011). A toxic characteristic leachate procedure test (TCLP) is needed to determine their viability for land disposal.

Spent brine will also concentrate other anions in the water. Complete characterization is necessary to ensure the arsenic concentration as well as the other ionic species are compatible with the requirements of the disposal site.

6.5 Pilot Testing

Chapter 13 of this handbook discusses the general process of piloting an ion exchange system. This section focuses on the specific considerations for arsenic removal. As with all bench and pilot testing, the test should mimic and be scalable to the anticipated full-scale system. Bench-scale testing can help determine the optimal empty bed contact time (EBCT) to conduct subsequent experiments, screen resins to identify best performing options, and demonstrate treatment performance through multiple loading and regeneration cycles.

A comprehensive water quality analysis should be performed that includes all of the anions and cations. Chapter 2 presents a table describing the typical water quality analyses. With respect to arsenic, both total arsenic, As(+3) + As(+5). Additional water parameters that are recommended include:

- pH,
- hardness,
- silica,
- manganese,
- sulfate,
- nitrate,
- phosphate,
- chloride,
- fluoride,
- antimony,
- total chromium,
- lead,
- molybdenum,
- total suspended solids,
- total iron,
- vanadium,
- total dissolved solids (TDS), and
- turbidity.

(USEPA 2005)

Based on manufacturer modeling, a pilot study may be beneficial to provide full-scale design parameters, including surface loading rate, throughput, EBCT, and waste characterization. The pilot column tests generally have multiple trains that allow the comparison of different resin types.

Occasionally, nitrate and arsenic can be found together. A typical SBA will preferentially remove arsenic, and to prevent chromatographic peaking, nitrate should be monitored well before theoretical breakthrough.

6.6 Design Guidelines

Recommended design guidelines for regenerable resins are further described here.

Pretreatment:

- Speciation of the forms of arsenic is recommended to determine if As(+3) and/or As(+5) are present and in what concentrations. If arsenic is present in the As(+3) form, it may need to be oxidized to As(+5). This is commonly achieved with chlorine. If disinfection byproducts (DBPs) are a concern or if chlorine residual can be damaging to the resin, consideration should be given to alternate oxidants.
- Prefilters to prevent total suspended solids (TSS) accumulation in the resin bed is vital to satisfactory operation. The resin will act as an absolute filter, producing water with an effluent turbidity of less than 0.3 NTU. The longer the resin will be in service, especially, the more important proper design and operation of a prefiltration system is to the resin performance. A particle size distribution analysis can help determine the size of incoming solids. If the well produces any sand, a de-sander is also recommended. The feed turbidity to the system should be less than 1.0 NTU.
- Resin performs best when foulants are eliminated. If any of these are present in the water in concentrations high enough to affect resin performance, they need to be adequately removed:
  - Iron and manganese >1 ppm
  - Oil and grease—any concentration is detrimental
  - Oxidants—any concentration is detrimental
  - Langelier Saturation Index—ensure there is no scaling potential, ≤0
  - Total organic carbon (TOC) >2 ppm

The recommended design guidelines for arsenic removal are

SBA design parameters:

- Resin: SBA
- Resin depth: 3 ft minimum (0.91 m)
- Hydraulic loading rate: 8–12 gpm/ft^2 (20–30 m/h)
- Specific flow rate: 1–5 gpm/ft^3 (8–40 BV/h)
- EBCT: 2–3 min minimum

If regenerating SBA resin:

- Regenerant solution: Sodium chloride brine
- Regenerant concentration: 6–10 percent NaCl
- Regenerant dosage: approximately 8–10 lb NaCl/ft^3 of resin

Regeneration steps:

- Backwash (as needed): Use manufacturer's backwash curves to achieve 50 percent bed expansion for anywhere from 1 to 3 BV depending on solids loading
- Brine step with regenerant solution: 1–2 BV of 6–10 percent NaCl at 0.25–0.5 gpm/ft^3 (2–4 BV/h)
- Slow rinse with treated water: typically, 0.25–0.5 gpm/ft^3 (2–4 BV/h) for 30–40 min
- Fast rinse with treated water: typically, at the service flow rate for 10–20 min

A key component in the design of ion exchange systems for arsenic removal is the estimation of the amount of water that can be treated before the resin is exhausted (referred to as throughput). The throughput is based on the water quality parameters and is generally obtained by

- evaluation of similar systems,
- estimates provided by resin suppliers based on modeling, and
- pilot testing.

6.7 Design Example

6.7.1 SBA Regenerable Example

In this example, the well has a capacity of 950 gpm and the influent arsenic is 30 ppb. The goal is to treat the water to ≤5 ppb as this will allow a factor of safety below the 10 ppb MCL. The sulfate level is 100 ppm, so SBA regenerable is a good option. Table 6-2 summarizes the raw water quality.

Table 6-2 SBA example raw groundwater parameters

Raw Water Parameter	Value
Well pump flow rate	950 gpm
Total arsenic	30 ppb
Alkalinity as $CaCO_3$	120 ppm as $CaCO_3$
Chloride	30 ppm
Sulfate	100 ppm
Nitrate as N	2 ppm as N
pH	7.2

Table 6-3 SBA example design parameters

Design Parameters	Value
Total flow rate	950 gpm
Influent arsenic	30 ppb
Effluent concentration goal	5 ppb
Number of vessels in service	1
Number of vessels in standby	1
Number of vessels total	2
Flow per vessel	950 gpm
Vessel diameter	10 ft
Volume per supersack	42 ft^3
Number of supersacks per vessel	6
Media volume per vessel	252 ft^3
Vessel area	78.5 ft^2
Bed depth per vessel	3.2 ft
Linear velocity: 8 to 12 gpm/ft^2 target per vessel	12 gpm/ft^2
Specific flow rate: 1 to 5 gpm/ft^3 target per vessel	2.8 gpm/ft^3
EBCT	2.6 minutes
BV/hour	23
Estimated throughput, BV	500 BVs
Bed life	21.7 hours
Backwash linear velocity	2.4 gpm/ft^2

BV: bed volume, EBCT: empty bed contact time

Table 6-3 provides a summary of the design parameters and the derivation of these parameters is described below.

A throughput projection can be done by contacting the resin supplier. In the example below, a throughput 500 BV was established based on a pilot test. From experience, the regenerable SBA resin estimated arsenate leakage would be about 2–3 ppb. Due to health effects, bypass is not recommended.

Treated flow = 950 gpm

Method of operation: Use one duty vessel and one in standby for regeneration.

Design Steps:

Vessel sizing

1. Choose the smallest commercially available standard diameter vessel that will provide sufficient cross-sectional area to produce a linear velocity in the target range of 8–12 gpm/ft^2. The following table presents commercially available vessel diameter with the cross-sectional area.

Commercial Vessel Diameter (ID) ft	Approx. Cross-sectional Area (ft^2)
12	113
11	95
10	78.5
9	63.6
8	50
7	38.5
6	28.3
5	19.6
4	12.6

Determine available equipment diameter using the linear velocity criteria:

- Assume linear velocity: 12 gpm/ft^2
- Vessel area = treated flow/linear velocity

$$\frac{950 \text{ gpm}}{12 \text{ gpm/ft}^2} = 79.2 \text{ ft}^2$$

In this case, a 10 ft diameter vessel will be chosen.

<u>Calculate resin volume</u>

For standard counter-flow regeneration, a minimum bed depth of 4 ft is needed. For a 10 ft diameter vessel, the minimum resin volume needed is as follows:

- Cross-sectional area of 78.5 ft^2 × 4 ft bed depth = 314 ft^3 approximately
- Since resin deliveries are usually in supersacks of 42 ft^3 each, the nearest whole number of supersacks should be chosen: 8 supersacks × 42 ft^3 = 336 ft^3
- Again, recheck resin bed depth: 336 ft^3/78.5 ft^2 = 4.3 ft—this is acceptable.

Determine linear velocity:
Linear velocity = flow (gpm) / resin vessel area (ft^2)
Linear velocity = 950 gpm / 78.5 ft^3 = 12 gpm/ft^2

Determine specific flow rate:
Specific flow rate = flow (gpm) / resin volume (ft^3)
Specific flow rate – 950 gpm / 336 ft^3 = 2.8 gpm/ft^3

Determine EBCT
EBCT = vessel volume/treated flow rate

$$336 \text{ ft}^3 \times \frac{7.48 \text{ gal}}{\text{ft}^3} \times \frac{\text{min}}{950 \text{ gal}} = 2.6 \text{ min}$$

2 min is within the EBCT goal criteria.

Calculate the days between exchanges:

Estimated throughput as provided by resin manufacturer = 500 BV

500 BV / 23 BV/h = 21.7 h

Design regeneration protocol

Regeneration steps:

- Backwash—typically expands the resin bed by 50–60 percent for 10 min (see note below)
- Regeneration—either co-flow or counter-flow
- Slow rinse—typically 7.5–15 gal/ft³ of resin (1–2 BV)
- Fast rinse—typically 23–38 gal/ft³ of resin (3–5 BV)

Backwash Note: For counter-flow regeneration, backwashing may be done infrequently, say once per month. Counter-flow backwashing has the potential to cause unintentional mixing of the highly regenerated resin at the outlet of the resin bed with the rest of the resin that is less so. This can lead to potentially higher than normal nitrate leakage during the following cycle. When backwashing is eventually done, the resin bed would usually be double regenerated before returning to service. During backwashing, the resin bed is generally expanded by 50–60 percent of its original settled volume to remove as much of the suspended solids that may have accumulated on the resin beads during the previous service period. It is important to get advice from the resin supplier on relevant backwash linear velocity rates that will accomplish the desired degree of bed expansion at the specific temperature of the water. A typical backwash rate at 50°F (10°C) can be 2 gpm/ft² (5 m/h). For a 10 ft (3.048 m) diameter vessel with a cross-sectional area of 78.5 ft², that equates to a flow rate of 157 gpm (35.6 m³/h). Usually, backwashing is done for at least 10–15 min; if the backwash effluent water is not relatively clear of suspended solids after this period, the backwash can be continued until effluent quality is satisfactory.

Brine Makeup and Application: A saturated brine solution (typically 26 percent) is usually prepared using sodium chloride salt that meets acceptable quality standards (e.g., ANSI/AWWA B200-17). Saturated brine contains approximately 2.6 lb of salt per gallon (approx. 300 grams per liter). During regeneration, the saturated brine solution is mixed with enough dilution water to create the desired concentration of brine (typically 5–12 percent) before it is passed through the resin bed. This roughly requires a dilution ratio of 2–1 for dilution water vs saturated brine. At 1 percent brine concentration, the salt content of each gallon of 10 percent brine will be approximately 0.9 lb.

Volume and Flowrate of 10 Percent Brine Solution: In this example, assume that the salt dosage applied is 10 lb salt per 1 ft³ of resin (or 160 g/L). Using a volume of 336 ft³ of resin in co-flow regeneration mode, the total volume of 10 percent brine needed can be calculated from:

$$336 \text{ ft}^3 \text{ of resin} \times \frac{10 \text{ lb of salt}}{\text{ft}^3 \text{ of resin}} = 3{,}360 \text{ lb of salt}$$

Therefore, the volume of 10 percent brine needed is:

$$\text{Vol of 10\% brine} = 3{,}360 \text{ lb of salt} \times \frac{1 \text{ gal}}{0.9 \text{ lb of salt}} = 3{,}733 \text{ gal of 10 percent salt}$$

Brine is usually applied at a specific flow rate of 0.25–0.5 gpm/ft^3 (2–4 BV/h) – the more contact time the better the efficiency. Ideally, the contact time between the brine and the resin should be a minimum of 30 min.

At 0.25 gpm/ft^3 and a volume of 3,733 gal of 10 percent brine, the brining time will be:

$$\frac{3{,}733 \text{ gal}}{336 \text{ ft}^3} \times \frac{\text{ft}^3}{0.25 \text{ gpm}} = 45 \text{ min (approx.)} \rightarrow \frac{3{,}733 \text{ gal}}{45 \text{ min}} = 84 \text{ gpm (approx.)}$$

Following the brine, the resin is usually rinsed to rid it of residual brine. Usually, the resin must be rinsed until the chloride content of the rinse water is about 25 ppm or lower compared to the chloride level in the raw water. At that point, the resin bed can be put into service. The first phase of rinsing, known as the slow rinse step, is conducted at the same flow rate used during the brining step. The objective of this is to continue the regeneration process of the resin by pushing out the brine in plug-flow fashion and ensuring that the nitrate and other anions are adequately removed from the resin in preparation for the next cycle.

For the slow rinse, 7.5–15 gal of water per ft^3 of resin is typically used.

Slow rinse water usage:

$$336 \text{ ft}^3 \text{ of resin} \times \frac{15 \text{ gal of water}}{\text{ft}^3 \text{ of resin}} = 5{,}040 \text{ gal of water}$$

Using the same 0.25 gpm/ft^3 specific flow rate used for the brining step, the slow rinse will be for 60 min and the flow rate will be 84 gpm.

$$\frac{5{,}040 \text{ gal}}{336 \text{ ft}^3} \times \frac{\text{ft}^3}{0.25 \text{ gpm}} = 60 \text{ min} \rightarrow \frac{5{,}040 \text{ gal}}{60 \text{ min}} = 84 \text{ gpm}$$

For the fast rinse step, the service flow rate can be used to save time. In this case, the service flow rate is 950 gpm, so the estimated time for the fast rinse, assuming 38 gal/ft^3 (5 BV) will be approximately:

$$336 \text{ ft}^3 \times \frac{38 \text{ gal}}{\text{ft}^3} \div 950 \text{ gpm} = 13.5 \text{ min (approx.)}$$

These estimates are summarized in Table 6-4.

Table 6-4 Regeneration: based on 10 lb/ft^3 NaCl and 10 percent brine solution

	Design Basis	Flow (gpm)	Time (min)	Total Waste per Step (gallons)
Backwash	2 gpm/ft^2	157	10	1,570
Brine step	0.25 gpm/ft^3	84	45	3,733
Slow rinse	0.25 gpm/ft^3	84	60	5,040
Fast Rinse		950	13.5	12,825
Totals			128.5	23,168

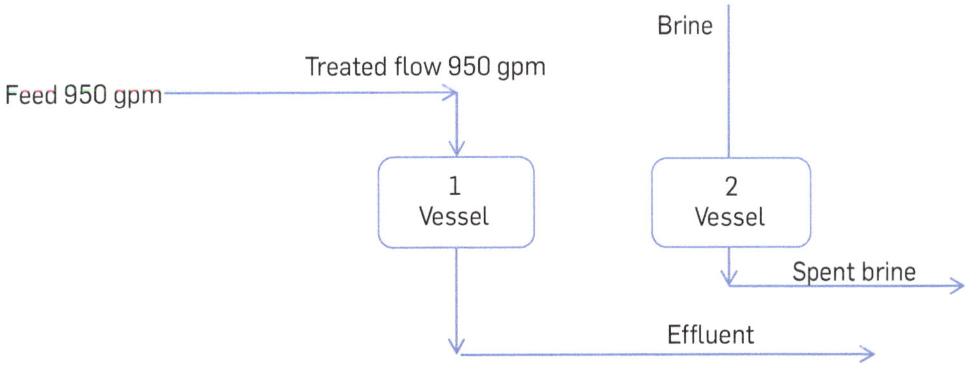

Figure 6-3 SBA regeneration co-flow regeneration example: Vessel 2 is offline while in regeneration

Courtesy of Purolite Corporation

Figure 6-3 provides a basic schematic of the two vessel ion exchange configuration for arsenic removal.

REFERENCES

Ahmad, S.A.; Khan, M.H.; and Haque, M. 2018. "Arsenic Contamination in Groundwater in Bangladesh: Implications and Challenges for Healthcare Policy." *Risk Management and Healthcare Policy* 11 (November): 251–61. https://doi.org/10.2147/RMHP.S153188.

Edzwald, J.K. 2011. *Water Quality and Treatment: A Handbook on Drinking Water, Sixth Edition*. 6th ed. AWWA, American Society of Civil Engineers, McGraw-Hill. https://www.awwa.org/Store/{ProductName}/ProductDetail/{ProductId}?productId=6292.

Health Canada. 2006. "Guideline Technical Document – Arsenic." Health Canada. https://www.canada.ca/en/health-canada/services/publications/healthy-living/guidelines-canadian-drinking-water-quality-guideline-technical-document-arsenic/page-4-guidelines-canadian-drinking-water-quality-guideline-technical-document-arsenic.html.

Loebenstein, J.R. 1994. "Materials Flow of Arsenic in the United States." Report 9382. Information Circular. USGS. USGS Publications Warehouse. http://pubs.er.usgs.gov/publication/70005343.

NTP (National Toxicology Program). 2016. *Report on Carcinogens*. 14th ed. U.S. Department of Health and Human Services. https://ntp.niehs.nih.gov/go/roc14.

Purolite. 2015. "Application Guide: Choosing an Ion Exchange System for Nitrate Removal." Applicaiton Guide P-000042-200PP-0915-R3-PCO. Applicaiton Guide. https://www.purolite.com/dam/jcr:5cb857fb-b8a1-46fd-8a1c-163042265086/Application%20Guide_Nitrate%20Removal_2015_Complete_09%202015_AC.pdf.

Sorg, T. and Lytle, D. 2019. "Treatment Options, Part 1." Water Suppy and Water Resources Division, Office of Research and Development, September. https://www.epa.gov/sites/production/files/2015-09/documents/treatment_options_tom_sorg.pdf.

USEPA (US Environmental Protection Agency). 2015a. "Arsenic Rule Compliance for Community Water System Owners and Operators." Collections and Lists. USEPA. September 23, 2015. https://www.epa.gov/dwreginfo/arsenic-rule-compliance-community-water-system-owners-and-operators.

USEPA. 2015b. "Drinking Water Arsenic Rule History." Collections and Lists. USEPA. October 13, 2015. https://www.epa.gov/dwreginfo/drinking-water-arsenic-rule-history.

USEPA. 2007. "Drinking Water Treatability Database." Drinking Water Treatability Database. March 2007. https://tdb.epa.gov/tdb/contaminant?id=1175876466.

USEPA. 2005. "Evaluating Arsenic Treatment Providers: A Guide for Public Water Systems." EPA-816-R-05-008. Office of Water. https://nepis.epa.gov/Exe/ZyPDF.cgi?Dockey=20017JUO.txt.

USEPA. 2001. "Technical Fact Sheet: Final Rule for Arsenic in Drinking Water." Technical Fact Sheet EPA 815-F-00-016. Environmental Protection Agency. https://nepis.epa.gov/Exe/ZyPdf.cgi?Dockey=20001XXE.txt.

WHO (World Health Organization). 1993. "Background Document for the Development of WHO Guidelines for Drinking-Water Quality.–2nded." *Volume* 2: 281–83.

CHAPTER 7
Nitrate Removal

7.1 Introduction

Nitrate (NO_3^-) occurs naturally in the environment, forming part of the nitrogen cycle in which atmospheric nitrogen is converted to ammonia (NH_3), nitrite (NO_2^-), and nitrate. These sources of nitrogen are vital for all plant life, being used to form amino acids and proteins. Naturally occurring nitrate concentrations are typically too low to pose drinking water health concerns, however human activities such as the application of nitrogenous fertilizers to arable land can significantly increase soil and water nitrate concentrations. Sewage treatment works discharges and other industrial sources can also contribute to elevated nitrate concentrations which may exceed safe drinking water standards.

Nitrates have no detectable color, taste, or odor at concentrations found in drinking water supplies, nor do they cause discoloration of plumbing fixtures, consequently they are undetectable to human senses (Desilva 2003). However, methemoglobinemia or "blue-baby" syndrome in bottle fed infants is a potentially fatal condition which may occur when nitrates are reduced to nitrites in the baby's stomach and combine with hemoglobin to form methemoglobin. This inhibits the blood's capacity to carry oxygen in the red blood cells (WHO n.d.). Some children and adults who ingest unusually high levels of nitrate may experience decreases in blood pressure, increased heart rate, reduced ability of the blood to carry oxygen to tissues, headaches, abdominal cramps, vomiting, and even death (Agency for Toxic Substances and Disease Registry 2017).

Treatment techniques such as reverse osmosis (RO), biological denitrification and electrodialysis can be used to remove nitrate from drinking water, although these generally tend to be less cost effective than ion exchange. Blending of different source waters containing high and low nitrate concentrations can be a cost-effective means of producing acceptable nitrate concentrations in drinking water however this option is not always available and generally is subject to additional regulatory scrutiny. In some instances, drilling a new borehole in a zone with lower nitrate concentrations may be the most appropriate solution. Conventional water treatment processes such as coagulation, clarification, filtration, activated carbon adsorption, and ultrafiltration membranes are not suitable for nitrate removal. Ion exchange water softeners are also unsuitable as these target positively charged cations rather than negatively charged nitrate anions (Woodward 2020). This chapter describes how ion exchange systems can be used for nitrate removal in potable water applications. This chapter describes:

- properties and chemistry associated with nitrate,
- regulatory aspects associated with nitrate,
- source water quality parameters that can impact the selection of ion exchange resins suitable for nitrate removal,
- an overview of bench and pilot testing requirements, and
- design guidelines including calculations and residuals generation
- design example.

7.2 Nitrate Occurrence and Chemistry

7.2.1 Occurrence and Uses

Nitrate occurs naturally in the environment. Nitrogen in the atmosphere is converted to ammonia (NH_3) by nitrogen-fixing bacteria in root nodules. Ammonia is oxidized to nitrite (NO_2^-) and nitrate (NO_3^-) by nitrifying bacteria in the soil. Nitrate is the final and most stable stage of ammonia and nitrite oxidation. Naturally occurring concentrations of nitrate in surface water are typically less than 2 mg/L (Mueller et al. 1995) which is well below the WHO Guideline Value of 50 mg/L as NO_3.

Nitrate concentrations in natural waters may be greatly increased by human activities, the most significant of which is the application nitrogenous fertilizers to farmland. Organic nitrogenous fertilizers include slurry, poultry manures, solid manures (such as farmyard manure, sludge cake or compost), sewage sludge (also called biosolids), and other liquid manures (such as abattoir waste or anaerobic digestate) (UK Government 2018). Human activity also contributes nitrate to the environment by discharges from municipal sewage treatment works, oxidation of nitrogenous waste from septic tanks, discharges from unlined landfill sites and other industrial sources.

Surface water nitrate concentrations have gradually increased in many European countries in the last few decades. In the United Kingdom, for example, rising trends in nitrate concentrations have been observed at many ground water abstraction points. According to work published by the British Geological Survey, the average trend across the UK was an increase of 0.34 mg/L as NO_3 per year (BGS 2020b).

Surface water nitrate concentrations can also be variable, with elevated concentrations usually attributable to increased runoff following rainfall and reduced biological activity in colder months. Groundwater nitrate concentrations generally change slowly over time, although (WHO 2017) seasonal variations can occur in winter and spring, corresponding to high water levels (BGS 2020a).

7.2.2 Chemistry

Most nitrogenous materials in natural waters such as ammonia and nitrite are eventually converted to nitrate as this is the most stable oxidation state (EPA). The nitrate ion is monovalent with a molecular weight of 62 g/mol.

Nitrates are soluble in water and are thus mobile in soil with a high potential to migrate to ground water. Nitrates do not volatilize and therefore are likely to remain soluble in water unless consumed by plants or other organisms (Rosenblatt 2005).

7.3 Nitrate Regulations

Elevated drinking water nitrate concentrations represent a public health risk particularly to young infants. Nitrate is regulated by the US Environmental Protection Agency (USEPA) and covered in guidelines provided by the World Health Organization (WHO). Nitrate concentrations are commonly given as nitrogen 'N' or as 'nitrate NO_3^-' where 1.0 mg/L nitrate as N is equivalent to 4.43 mg/L nitrate as NO_3^- corresponding to the ratio of molecular weights of nitrogen (14 g/mol) and nitrate (62 g/mol) respectively.

7.3.1 EPA Regulations

The EPA has established a Maximum Contaminant Level (MCL) of 10 mg/L as N which is equivalent to 44.3 mg/L as NO_3^-. This limit was set to protect infants below the age of six months who can suffer from methemoglobinemia or blue baby syndrome if they consume water with high nitrate (USEPA 2013).

7.3.2 International Regulations and Guidelines

WHO does not have specific regulations but rather guidelines. In its 2017 document WHO Guidelines for Drinking-water Quality (Fourth Edition, incorporating the first addendum), WHO provided a guideline of 50 mg/L as NO_3^-, which is equivalent to 11.3 mg/L as N (WHO 2011).

Health Canada provides Guidelines for Canadian Drinking Water Quality. Health Canada set a Maximum Acceptable Concentration (MAC) of 45 mg/L as NO_3^-, which is equivalent to approximately 10 mg/L as N. According to Health Canada, this MAC is protective of the health of the most sensitive sub-population, bottle-fed infants (Canada 1997).

The European Union Directive 98/83/EC nitrate parametric value is 50 mg/L as NO_3^-. Member States must also ensure that [nitrate]/50 + [nitrite]/3 $\leq$ 1, the square brackets signifying the concentrations in mg/L for nitrate and nitrite. This is to account for the possibility of the simultaneous occurrence of nitrate and nitrite in drinking-water.

7.4 Ion Exchange Resin Treatment Options for Nitrate

A detailed raw water analysis is required to aid ion exchange resin selection. The following, constituents should be considered: calcium, magnesium, sodium, potassium, iron, manganese, sulfate, nitrate, bicarbonate, chloride, bromide, fluoride, arsenic (As[V]), hexavalent chromium, uranium, total dissolved solids (TDS), conductivity, total suspended solids (TSS), total organic carbon (TOC), turbidity, pH, and water temperature. Current and future design horizon values should be considered.

It is typical in larger plants to bypass a portion of the water around the ion exchange process, producing a blend of treated and untreated water that meets the maximum allowable nitrate level in drinking water. A mass balance should be carried out to assess the proportion of the water to be treated by ion exchange versus the amount to be bypassed.

7.4.1 Nitrate Removal Ion Exchange Resins

Strong base anion (SBA) resins in the chloride form are typically used for drinking water nitrate removal applications. Nitrate and other anions are exchanged for the chloride ion (Cl^-) with a typical ion exchange reaction as follows:

$$NO_3^- + \text{Resin-}Cl^- \leftrightarrow \text{Resin-}NO_3^- + Cl^-$$

The dual arrows indicate that the process is reversible with the resin being regenerated using a sodium chloride brine solution.

Resins available for nitrate removal are of three main types (Purolite 2015):

- Type 1 SBA resin
- Type 2 SBA resin
- Nitrate selective SBA resin

These resins are effective over wide ranges of temperatures and pH. Weak base anion (WBA) resin can also be used (e.g., Purolite patented process), but the resin must be protonated before it can remove nitrate and other anions—this usually requires pH reduction of the feedwater. WBA resins are reserved for treating waters with very high levels of nitrate such as in agricultural run-offs and as such will not be discussed further. Designers should refer to local drinking water approved product lists before selecting an ion exchange resin. Major differences in these three types of resins are discussed below.

7.4.2 Comparing Type 1, Type 2, and Nitrate Selective Strong Base Resins

Type 1 SBA ion exchange resins are manufactured with trimethylamine functional groups, while Type 2 resins are manufactured with dimethylethanolamine functional groups.

One drawback of Type 1 SBA resins is the tendency to give off a slight fishy amine smell that may be detected by some sensitive customers even at very low parts per trillion concentrations (Purolite 2015). The phenomenon is particularly noticeable in systems that experience frequent stop/start operation such as for small community water systems or point-of-entry (POE) household use. Therefore Type 1 resins are generally not recommended for such service (Purolite 2015). Type 1 resins are however acceptable for use municipal plants since these plants tend to operate continuously and the long distribution lines to consumers help to dissipate odor. Another good reason to consider Type 1 resins for municipal use is the availability in the last decade or so of higher capacity versions that tend to rival the economics of the other types of resins. More information on Type 1 resins is provided in Chapter 2: Fundamentals of Ion Exchange.

Type 2 SBA resins do not generally impart odor to the water and can be used across the entire spectrum of the municipal, small community, and POE systems.

For both Type 1 and Type 2 resins, the relative order of affinity for common anions is as follows:

$$SO_4^{2-} > HAsO_4^{2-} > NO_3^- > Cl^- > HCO_3^- > F^-.$$

Type 1 and Type 2 SBA resins have a higher affinity for sulfate and arsenic than for nitrate. Passage of water through the ion exchange column results in sulfate being preferentially loaded at the inlet of the column with nitrate and other anions progressively pushed deeper into the ion exchange column. When sulfate reaches the end of the bed it displaces the concentrated nitrate which can exit the column at a concentration higher than the influent concentration. This is known as chromatographic peaking (or dumping) of nitrate and is discussed more extensively in Chapter 2. Treatment plants are therefore designed to stop the service cycle at a safe nitrate breakthrough point so peaking does not occur. At that point, it is important to use the correct amount of salt to regenerate the resin in preparation for the next service cycle.

An example of nitrate peaking before sulfate breaks through is shown in Figure 7-1 for a Type 2 SBA resin. In this example, arsenic (As[V]) is also present in the water and peaks after nitrate but before sulfate breaks. Thus, the removal of both nitrate and arsenic can be safely achieved using the nitrate breakpoint for terminating the run.

Nitrate selective resins (sometimes referred to as nitrate-over-sulfate-selective [NSS] resins) were developed for treating relatively high sulfate water and to overcome the nitrate dumping or chromatographic peaking issue observed with Type 1 and Type 2 SBA resins (Clifford 1999). With nitrate selective resins, the ion selectivity is reversed for sulfate and nitrate compared to Type 1 and Type 2 SBA resins:

$$NO_3^- > SO_4^{2-} > HAsO_4^{2-} > Cl^- > HCO_3^- > F^-$$

Nitrate selective resins are similar to standard Type 1 SBA resins but are generally made using triethylamine functional groups instead of trimethylamine

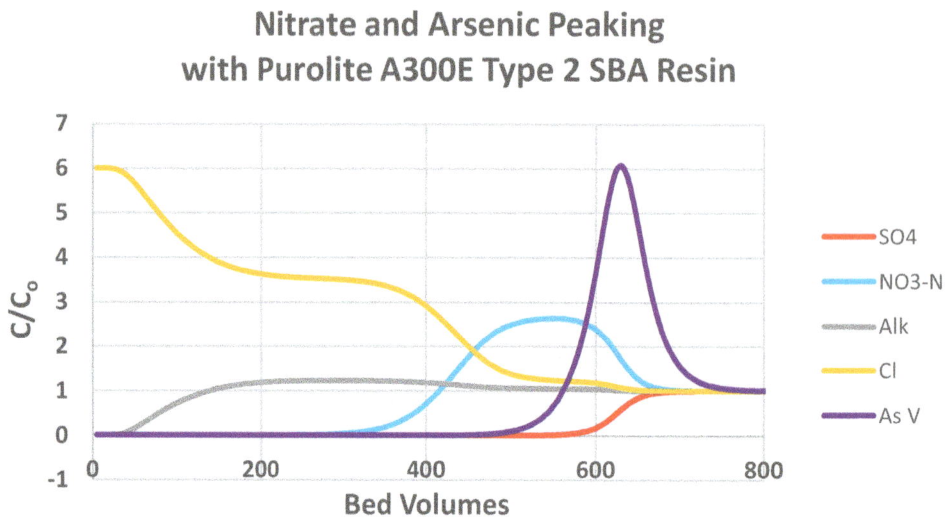

Figure 7-1 Nitrate peaks before arsenic with a Type 2 SBA resin
Courtesy of Purolite Corporation

as used for standard Type 1 resins. The higher molecular weight of the triethyl groups vs the trimethyl groups results in ion exchange sites that are more hydrophobic with higher affinity for hydrophobic ions like nitrate versus sulfate. Higher molecular weight functional groups made from tripropylamine and tributylamine show an even higher affinity for nitrate over sulfate as these provide increased charge-separation distance between functional sites (AWWA—*Water Quality and Treatment,* 6th edition). Field experience, however, shows that such resins are more difficult to regenerate and require higher levels for adequate regeneration, essentially countering any gains made from the higher nitrate affinity.

A traditional approach was to favor the use of nitrate selective resins over Type 1 and Type 2 resins when the ratio of sulfate to nitrate in the raw water exceeded 60:40. In recent years, however, higher capacity Type 1 SBA resins have been developed and these can rival and even exceed the operating capacity of nitrate selective resins, depending on water chemistry. Therefore, the blanket application of this old rule of thumb is no longer recommended. Instead, all resins should be considered for municipal treatment and selection should be made based on the resin that provides the best operating economics while not exceeding the target nitrate leakage in the treated water. Calculating the most cost-effective resin should consider the cost of the resin, the cost of regenerant per volume of water treated and the cost of wastewater disposal.

In cases where simultaneous removal of nitrate and arsenic is needed, nitrate selective resins should not be used as arsenic will peak before nitrate breaks as seen in Figure 7-2.

Since arsenic is more toxic than nitrate, it is preferable to use either a Type 1 or Type 2 SBA resin as nitrate breaks before arsenic. Because nitrate dumping is not experienced with nitrate selective resins, they are especially useful for POE

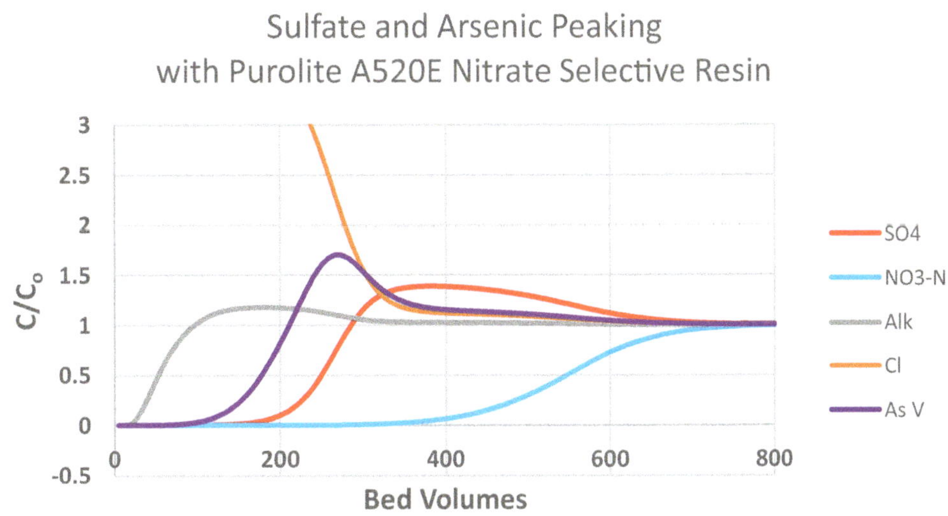

Figure 7-2 No nitrate dumping with a nitrate selective resin
Courtesy of Purolite Corporation

household systems where the possibility is higher for system malfunction with the risk of possible overruns which can go unnoticed for some time.

7.4.3 Continuous Ion Exchange Systems

Continuous ion exchange systems such as the Calgon ISEP® system or the ACWA Services Nitreat® system are intended to reduce regeneration brine requirements and minimize the volume of waste produced.

These systems are configured to be true counter-flow ion exchange systems comprising a series of small resin vessels connected via a central multiport valve or turntable. The central valve directs feedwater, treated water, regenerant brine and rinse water to the resin vessels, indexing around them at a set interval. In this way, all the vessels continually cycle through adsorption, regeneration, and rinse phases with the vessel containing the most exhausted resin moving out of the cascade for regeneration as a recently regenerated vessel enters into service at the other end (Calgon 2021).

7.5 Treated Water Corrosivity and Alkalinity

The removal of nitrate and other anions by SBA ion exchange resins operating in the chloride form increases the chloride content of the treated water. Depending on the total concentration of anions present in the feedwater, the initial chloride concentration can be generally two to five times higher than its influent value. Such changes can potentially increase the corrosivity of the water to downstream piping.

There is no WHO health-based guideline value for chloride in drinking-water, however, some countries may have secondary drinking water guidelines or regulatory limits (e.g., 250 mg/L in the United States and United Kingdom). Chloride concentrations above about 250 mg/L can give rise to detectable taste in water (WHO 1996).

The reduction in pH and alkalinity occurs as bicarbonate ions are exchanged for chloride ions. The combination of reduced pH and alkalinity may increase the corrosivity of the water as assessed by the Langelier Saturation Index (LSI) and will reduce the calcium carbonate precipitation potential (CCPP). The Larson Ratio (LR), which computes a corrosion index from the ratio of the sum of the concentrations of chloride (Cl^-) and sulfate (SO_4^{2-}) divided by the concentration of bicarbonate (HCO_3^-), will also show a higher potential for corrosion of downstream piping. If a bicarbonate wash step is included as part of the regeneration, reduction of pH and alkalinity reduction at the start of the service cycle can be largely eliminated, leading to milder potential for lead corrosion (Purolite). Refer to Chapter 12 on simultaneous compliance.

Designers must assess the effect of ion exchange as well as other processes to ensure the final water is stable with respect to pH, non-aggressive to concrete and non-corrosive to metals in the distribution network. In some instances, a bicarbonate wash step is added to partially convert the resin to a mixed chloride/bicarbonate form for return to service. This can help eliminate the pH drop that can be experienced at the start of the service cycle.

7.6 Regeneration Options

Regeneration of Type 1, Type 2, and nitrate selective SBA exchange resins is achieved by passing a solution of sodium chloride brine (typically at 5–12 percent) through the ion exchange resin. Chloride ions displace nitrate and other adsorbed ions, restoring the capacity of the ion exchange resin. The brine solution is normally held in salt saturators in which water is mixed with PDV salt (pure dried vacuum salt). The regeneration cycle is typically initiated on differential pressure or timed basis and a typical regeneration cycle consists of the following steps:

- Backwash raw water (largely to remove particulates and fluidize the bed)
- Co-flow or counter-flow regeneration with sodium chloride brine
- Displacement or slow rinse using softened water
- Optional bicarbonate wash to partially convert the resin to bicarbonate form to control effluent pH and alkalinity
- Fast rinse
- Return to service

Where the hardness of the water is greater than 170 mg/L as $CaCO_3$, softened water is required for the brine makeup and rinse and steps to prevent $CaCO_3$ scaling (Purolite). Water is typically softened by cation exchange resin (see Chapter 3: Hardness Removal).

Complete regeneration of nitrate removal ion exchange resin is not considered to be efficient in terms of the amount of salt required. Incomplete or partial regeneration (removal of 50–60 percent of the sorbed contaminant) is typically acceptable for nitrate removal applications since a controlled amount of post-regeneration nitrate leakage is permissible (typically up to 7 mg NO_3^-N per liter) (AWWA).

As discussed in Chapter 2, regeneration may be:

a. co-flow, where both the water to be treated and the brine regenerant pass through the ion exchange bed in the same direction, generally downflow, or
b. counter-flow, where the water to be treated and the brine regenerant pass through the ion exchange bed in opposite directions (Purolite 2015).

With co-flow regenerated systems, sulfate and nitrate tend to load on the resin bed predominantly at the inlet end of the bed, whereas chloride and bicarbonate tend to load largely at the outlet end. When brine is introduced for regeneration at the inlet of the bed, sulfate and nitrate are stripped from the inlet end of the bed and pushed toward the outlet end of the bed. Consequently, some nitrate ions exchange with chloride and bicarbonate ions toward the outlet end. When the resin bed is returned to service, some of the nitrate bound to the resin at the exit of the resin bed is released and contributes to an initial elevated level of nitrate leakage (Purolite 2015).

Counter-flow regeneration reduces nitrate leakage on startup as the feedwater and the regenerant brine enter and leave the resin bed in opposite directions. Nitrate loaded on the ion exchange resin during service is pushed out of the bed without moving through the entire bed during regeneration.

7.7 Wastewater Disposal Options

Wastewater to be disposed of includes the spent regenerant brine, rinse and backwash water. In more advanced systems, it may be possible to recover some of the spent regenerant, rinse or backwash waters to reduce waste production.

When combined with the spent regenerant, rinse and backwash water result in a more dilute brine than the typical starting concentration of the fresh brine. The spent brine consists largely of sodium salts of residual chloride and sulfate, nitrate and bicarbonate eluted from the resin. The wastewater volume typically ranges from about 3 percent to as low as 0.2 percent, depending on the chemistry of the water being treated (Purolite 2015).

Direct discharge of the waste volume to the environment is rarely permitted. Options for disposal include disposal to a local wastewater treatment plant if permitted, or haulage to a larger wastewater plant where it can be sufficiently diluted with a larger volume of wastewater before treatment. Road transportation of wastewater to a sewage treatment works can be a major contributor to the overall treatment works operational costs. In rare cases, deep well injection may be permitted.

The sulfate concentration in regenerant wastewater can be very significant (several thousand mg/L). Sulfate attack is therefore a risk in concrete and mortar structures should these receive the wastewater. Concrete tanks should therefore be avoided, or measures taken to coat or protect them appropriately against sulfate attack.

7.8 Resin Replacement Frequency

It is not unusual for resins used in nitrate removal applications to last 10–15 years or longer before replacement is needed (Purolite). Physical and chemical degradation of the resin beads in this type of application is usually a slow process. Except for dissolved oxygen, there are generally no other oxidants present in the feedwater that can chemically degrade the resin. The exchange of sulfate and nitrate for chloride by the resin results in minimal changes in the volume of the resin, resulting in minimal mechanical stress from the mild expansion and contraction that occurs from cycle to cycle. Backwash, when performed correctly, subjects the resin beads to minimal physical forces, resulting in minimal breakage of the beads. The minor resin losses that do occur are usually due to broken beads or loss of some whole beads during the backwash step. Usually, no more than 1–2 percent of the volume needs to be topped up annually.

7.9 Bench and Pilot Testing

Chapter 13 of this handbook discusses the general process of piloting ion exchange systems. This section focuses on the specific considerations for nitrate removal. Nitrate removal is an established process and resin suppliers can usually provide fairly accurate estimates of operating capacity and nitrate leakage based on influent water quality. This usually precludes the need to pilot. Piloting is generally reserved for unusual or extreme cases.

If piloting is required, as with all bench and pilot testing, the test should mimic and be scalable to the anticipated full-scale system. Bench-scale testing can help screen resins to identify best performing options and demonstrate treatment performance through multiple loading and regeneration cycles. Bench-scale pilot testing, if done properly, can be used to go directly to full-scale design. In such a case, the resin column bed depth and linear velocity used should preferably match the values expected to be used in the full-scale plant. A minimum column diameter of four inches is recommended to avoid the potential for wall effects that can be experienced with smaller diameter columns.

A comprehensive water quality analysis should be performed that includes all of the anions and cations. Chapter 2 presents a table describing the typical water quality analyses needed that includes the following:

- Calcium
- Magnesium
- Sodium
- Potassium
- Total iron
- Manganese
- Sulfate
- Nitrate
- Bicarbonate alkalinity
- Chloride
- Bromide
- Fluoride
- Arsenic (As[III] and As[V])
- Total chromium (Cr[III] and Cr[VI])
- Uranium
- TOC
- TDS/conductivity
- TSS
- Turbidity
- pH
- Temperature

7.10 Design Guidelines

Recommended design guidelines for regenerable resins are described here. The brine regenerant concentration and dosage can vary depending on the raw water contaminant level. The higher the contaminant level, the higher the concentration and dosage.

Ion exchange resin beds typically achieve treated water nitrate concentrations of approximately 1.0 mg/L as N or approximately 5 mg/L as NO_3. This is significantly lower than drinking water maximum concentration limits such as the EPA MCL of 10 mg/L as N. Therefore, ion exchange systems normally treat a proportion of the flow and blend the treated water with flows that bypass the ion exchange plant. A mass balance approach is used to determine the ratio of treated and bypassed flows needed to achieve the required final water nitrate concentration. In the United Kingdom, typical target final water nitrate concentration is approximately 42 mg/L as NO_3, against a regulatory limit of 50 mg/L as NO_3. In operation, the ratio of treated and bypassed flows is controlled automatically to achieve the final water nitrate concentration set point.

The recommended design guidelines for fixed bed nitrate removal systems are as follows:

- Resin: SBA in the Cl^- form
- Linear velocity (LV): Typical: 8–12 gpm/ft^2 (approx. 20–30 m/h)

- Resin depth: 3 ft (0.91 m) minimum for co-flow regeneration, 4 ft. (1.2 m) minimum for counter-flow
- Specific flow rate: 1–5 gpm/ft^3 (8–40 bed volumes [BV]/h)
- Regenerant solution: sodium chloride brine
- Regenerant concentration: 5–12 percent NaCl
- Regenerant dosage: 7.5–12 lb/ft^3 (120–192 g/L)

Please see the included nitrate case study in this handbook for an example of a continuous system.

7.11 Design Example

A well with a capacity of 700 gpm contains nitrate above the MCL. Table 7-1 presents the raw water quality parameters.

The target nitrate concentration in the effluent from the resin is an average of 5 mg/L-NO_3 (1.1 mg/L as N).

1. Vessel diameter sizing: initially choose a linear velocity (LV) of 12 gpm/ft^2 (approximately 30 m/h)

Table 7-1 Raw groundwater parameters

Raw Water Parameter	Value
Well pump flow rate	700 gpm (159 m^3/h)
Firm treatment capacity	700 gpm (159 m^3/h)
Sewer availability	Yes
Nitrate	57.0 mg/L as NO_3
Sulfate	15.2 mg/L
Chloride	28.1 mg/L
Bicarbonate	156.9 mg/L as $CaCO_3$
Calcium	85.2 mg/L
Magnesium	6.0 mg/L
Sodium	63.61 mg/L
Total organic carbon	0.5 mg/L
Iron	0.1 mg/L
Manganese	0.05
Total hardness	120 mg/L as $CaCO_3$
Suspended solids	<0.2 mg/L
Conductivity	500 µS/cm
pH	7.7

Choosing an LV at or close to the upper end of the acceptable LV range will tend to reduce the size, cost, and the number of ion exchange vessels needed.

a. Calculate nearest vessel cross-sectional area needed:

$$700 \text{ gpm} / 12 \text{ gpm/ft}^2 = 58.33 \text{ ft}^2$$

b. Choose the smallest commercially available standard diameter vessel from the following table that meets the cross-sectional area requirement above:

Commercial Vessel Diameter (ID) ft	Approx. Cross-sectional area (ft^2)
12	113
11	95
10	78.5
9	63.6
8	50
7	38.5
6	28.3
5	19.6
4	12.6

In this case, either a 9 ft or 10 ft diameter vessel will be chosen, depending on availability. Since 10 ft diameter vessels are more readily available, this is chosen for the exercise.

2. Calculate resin volume needed:

For co-flow regeneration, a minimum resin bed depth of approximately 3 ft is recommended. For a 10 ft diameter vessel, the minimum resin volume needed is as follows:

Cross-sectional area of 78.5 ft^2 × 3 ft bed depth = 236 ft^3 approx.
Since resin deliveries are usually in supersacks of 42 ft^3 each, the nearest whole number of supersacks should be chosen: 6 supersacks × 42 ft^3 = 252 ft^3.
Then, recheck resin bed depth: 252 ft^3/78.5 ft^2 = 3.2 ft—this is acceptable.

For standard counter-flow regeneration, a minimum bed depth of 4 ft is needed. For a 10 ft diameter vessel, the minimum resin volume needed is as follows:

Cross-sectional area of 78.5 ft^2 × 4 ft bed depth = 314 ft^3 approx.
Since resin deliveries are usually in supersacks of 42 ft^3 each, the nearest whole number of supersacks should be chosen: 8 supersacks × 42 ft^3 = 336 ft^3.
Again, recheck resin bed depth: 336 ft^3/78.5 ft^2 = 4.3 ft—this is acceptable.

Table 7-2 Performance comparison of nitrate selective vs Type 1 resin

	Nitrate Selective	Type 1
	A520E	A600E/9149
Average Leakage		
Average nitrate leakage (ppm as NO_3)	4.81	4.44
Average nitrate leakage (ppm as N)	1.09	1.00
Operating Specifications		
Cycle time (h)	18.6	23.0
Net water per cycle, US gal/cycle	781,200	966,000
Regenerant		
Regeneration mode	co-flow	co-flow
Sodium chloride dosage, lb/ft^3	10.0	10.0
Regenerant concentration, percent	10.0	10.0
Total applied, lb	2,519.99	2,519.99
Total applied, equivalents	19,539.35	19,539.35
Excess (eq)	17,031.68	16,420.55
Regenerant as percent of theory	779.18	626.50
Capacity		
Available capacity, eq/L (kg/ft^3)	0.39 (8.51)	0.49 (10.63)
Design factor	0.90	0.90
Operating capacity, eq/L (kg/ft^3)	0.35 (7.68)	0.44 (9.55)

Courtesy of Purolite Corporation

Service capacity:

An example of a system design that can be provided by resin suppliers is shown in Table 7-2. The example compares a nitrate selective resin with a high capacity Type 1 SBA resin while using co-flow regeneration.

This example reinforces the need to consider all resin types when designing for municipal systems. The projection shows that the high capacity Type 1 SBA resin should provide higher operating capacity and consume 20 percent less salt while producing water with comparable nitrate leakage.

Regeneration steps:

- Backwash—typically expands the resin bed by 50–60 percent for 10 min (see note)
- Regeneration—either co-flow or counter-flow
- Slow rinse—typically 7.5–15 gal/ft^3 of resin (1–2 BV)
- Fast rinse—typically 23–38 gal/ft^3 of resin (3–5 BV)

Backwash note: The backwashing step is done as part of each regeneration procedure if the system is designed for co-flow regeneration. For counter-flow regeneration, backwashing may be done infrequently, about once per month. Counter-flow backwashing has the potential to cause unintentional mixing of the highly regenerated resin at the outlet of the resin bed with the rest of the resin that is less so. This can lead to potentially higher than normal nitrate leakage during the following cycle. When backwashing is eventually done, the resin bed would usually be double regenerated before returning to service. During backwashing, the resin bed is generally expanded by 50–60 percent of its original settled volume to remove as much of the suspended solids that may have accumulated on the resin beads during the previous service period. It is important to get advice from the resin supplier on relevant backwash linear velocity rates that will accomplish the desired degree of bed expansion at the specific temperature of the water. A typical backwash rate at 50°F (10°C) can be 2 gpm/ft^2 (5 m/h). For a 10 ft (3.048 m) diameter vessel with a cross-sectional area of 78.5 ft^2, that equates to a flow rate of 157 gpm (35.6 m^3/h). Usually, backwashing is done for at least 10–15 min; if the backwash effluent water is not relatively clear of suspended solids after this period, the backwash can be continued until effluent quality is satisfactory.

Brine makeup and application: A saturated brine solution (typically 26 percent) is usually prepared using sodium chloride salt that meets acceptable quality standards (e.g., ANSI/AWWA B200-17). Saturated brine contains approximately 2.6 lb of salt per gallon (approx. 300 g/L). During regeneration, the saturated brine solution is mixed with enough dilution water to create the desired concentration of brine (typically 5–12 percent) before it is passed through the resin bed. This roughly requires a dilution ratio of 2 to 1 for dilution water vs saturated brine. At 10 percent brine concentration, the salt content of each gallon of 10 percent brine will be approximately 0.9 lbs.

Volume and flow rate of 10 percent brine solution: In this example, assume that the salt dosage applied is 10 lb salt per 1 ft^3 of resin (160 g/L). Using a volume of 336 ft^3 of resin in co-flow regeneration mode, the total volume of 10 percent brine needed can be calculated from:

$$336 \text{ ft}^3 \text{ of resin} \times \frac{10 \text{ lb of salt}}{\text{ft}^3 \text{ of resin}} = 3{,}360 \text{ lb of salt}$$

Therefore, the volume of 10% brine needed is

$$\text{Vol of 10\% brine} = 3{,}360 \text{ lb of salt} \times \frac{1 \text{ gal}}{0.9 \text{ lb of salt}} = 3{,}733 \text{ gal of 10 percent salt}$$

Brine is usually applied at a specific flow rate of 0.25–0.5 gpm/ft^3 (2–4 BV/h)—the more contact time the better the efficiency. Ideally, the contact time between the brine and the resin should be a minimum of 30 min.

At 0.25 gpm/ft^3 and a volume of 3,733 gal of 10 percent brine, the brining time will be:

$$\frac{3{,}733 \text{ gal}}{336 \text{ ft}^3} \times \frac{\text{ft}^3}{0.25 \text{ gpm}} = 45 \text{ min (approx.)}$$

Following the brine, the resin is usually rinsed to rid it of residual brine. Usually, the resin must be rinsed until the chloride content of the rinse water is about 25 mg/L or lower compared to the chloride level in the raw water. At that point, the resin bed can be put into service. The first phase of rinsing, known as the slow rinse step, is conducted at the same flow rate used during the brining step. The objective of this is to continue the regeneration process of the resin by pushing out the brine in plug-flow fashion and ensuring that the nitrate and other anions are adequately removed from the resin in preparation for the next cycle.

For the slow rinse, 7.5–15 gal (28–57 L, approximately) of water per cubic feet of resin is typically used.

Slow rinse water usage:

$$336 \text{ ft}^3 \text{ of resin} \times \frac{15 \text{ gal of water}}{\text{ft}^3 \text{ of resin}} = 5{,}040 \text{ gal of water}$$

Using the same 0.25 gpm/ft^3 specific flow rate used for the brining step, the slow rinse flow rate will be 84 gpm, requiring a total of 60 min.

$$\frac{5{,}040 \text{ gal}}{336 \text{ ft}^3} \times \frac{\text{ft}^3}{0.25 \text{ gpm}} = 60 \text{ min}$$

For the fast rinse step, the service flow rate can be used to save time. In this case, the service flow rate is 700 gpm, so the estimated time for the fast rinse, assuming 38 gal/ft^3 (5 BV) will be approximately:

$$336 \text{ ft}^3 \times \frac{38 \text{ gal}}{\text{ft}^3} \div 700 \text{ gpm} = 18 \text{ min (approx.)}$$

These estimates are summarized in Table 7-3.

Table 7-3 Steps for regeneration

	Design Basis	Flow (gpm)	Time (min)	Total Waste per Step (gallons)
Backwash	2 gpm/ft^2	157	10	1,570
Brine step	0.25 gpm/ft^3	84	45	3,733
Slow rinse	0.25 gpm/ft^3	84	60	5,040
Fast Rinse		700	18	12,600
Totals			133	22,943

Counter-flow regeneration usually results in more efficient regeneration of the resin bed vs. co-flow regeneration. In the case of a counter-flow regenerated resin bed, the service flow would be in one direction through the bed while the brine and slow rinse will be in the opposite direction. This allows the freshest brine to make contact with the resin beads that are closest to the service outlet end of the vessel. The result is deeper cleaning of the resin in that area of the bed and generally lower nitrate leakage during the service phase. The trade-off is a more costly equipment design and a somewhat more complex method of operation.

Residuals handling: Total wastewater volume expected can be computed from the addition of the volumes of backwash, brine, and rinse. In this case, wastewater volume is approximately 22,943 gal (about 9 BV), or about 1.8 percent of the water treated. The composition of the spent brine will be a mixture of sodium salts comprised largely of chloride but also sulfate, nitrate and bicarbonate that was eluted from the resin.

REFERENCES

Agency for Toxic Substances and Disease Registry. 2017. "Toxicological Profile for Nitrate and Nitrite," July, 324.

BGS (British Geological Survey). 2020a. "Nitrate Contamination | Groundwater | British Geological Survey (BGS)." 2020a. https://www.bgs.ac.uk/research/groundwater/quality/nitrate/home.html.

BGS. 2020b. "Trends in Nitrate Concentrations in Groundwater | Groundwater | British Geological Survey (BGS)." British Geological Survey. 2020b. https://www.bgs.ac.uk/research/groundwater/quality/nitrate/trends.html.

Carbon Calgon. 2021. "ISEP." Calgon Carbon Corporation. January 2021. https://www.calgoncarbon.com/products/isep/.

Desilva, F. 2003. "Nitrate Removal by Ion Exchange." Water & Wastes Digest. March 27, 2003. https://www.wwdmag.com/nitrate-removal-ion-exchange.

Health Canada. 1997. "Page 2: Guidelines for Canadian Drinking Water Quality: Guideline Technical Document - Nitrate and Nitrite." Research;guidance. Aem. August 29, 1997. https://www.canada.ca/en/health-canada/services/publications/healthy-living/guidelines-canadian-drinking-water-quality-guideline-technical-document-nitrate-nitrite/page-2-guidelines-canadian-drinking-water-quality-guideline-technical-document-nitrate-nitrite.html.

Mueller, D.K.; Hanilton, P.A.; Helsel, D.R.; Hitt, K.J.; and Ruddy, B.C. 1995. "Nutrients in Ground Water and Surface Water of the United States-An Analysis of Data Through 1992" 95–4031. U.S. Geological Survey Water Resources Investigation Report. https://water.usgs.gov/nawqa/nutrients/pubs/wcp_v39_no12/.

Purolite. 2015. "Application Guide: Choosing an Ion Exchange System for Nitrate Removal." Applicaiton Guide P-000042-200PP-0915-R3-PCO. Applicaiton Guide. https://www.purolite.com/dam/jcr:5cb857fb-b8a1-46fd-8a1c-163042265086/Application%20Guide_Nitrate%20Removal_2015_Complete_09%202015_AC.pdf.

Rosenblatt, D. 2005. "Inert Ingredient Tolerance Reassessment - Ammonium Nitrate, Magnisium Nitrate, Sodium Nitrate, and Sotium Nitrite." Action Memorandum. EPA Office of Prevention, Pesticides and Toxic Substances. https://www.epa.gov/sites/production/files/2015-04/documents/nitrate_0.pdf.

UK Government. 2018. "Using Nitrogen Fertilisers in Nitrate Vulnerable Zones." Guidance. GOV.UK. August 2018. https://www.gov.uk/guidance/using-nitrogen-fertilisers-in-nitrate-vulnerable-zones.

USEPA (US Environmental Protection Agency). 2013. "Estimated Nitrate Concentrations in Groundwater Used for Drinking." Data and Tools. USEPA. March 27, 2013. https://www.epa.gov/nutrient-policy-data/estimated-nitrate-concentrations-groundwater-used-drinking.

WHO. 2017. "Guidelines for Drinking-Water Quality." In *Chemical Fact Sheet*, 4th ed., 307–442. Geneva: World Health Organization.

WHO. 2011. "Nitrate and Nitrite in Drinking-Water." Background document for development of WHO Guidelines for Drinking-water Quality. Geneva: World Health Organization. https://www.who.int/water_sanitation_health/dwq/chemicals/nitratenitrite2ndadd.pdf.

WHO. 1996. "Chloride in Drinking-Water." Background document for development of WHO Guidelines for Drinking-water Quality. Geneva: World Health Organization. https://www.who.int/water_sanitation_health/dwq/chloride.pdf.

WHO. "Water-Related Diseases." n.d. WHO. World Health Organization. Accessed July 13, 2020. http://www.who.int/water_sanitation_health/diseases-risks/diseases/methaemoglob/en/.

Woodward, J. 2020. "How to Remove Nitrates from Water." Fresh Water Systems. January 7, 2020. https://www.freshwatersystems.com/blogs/blog/how-to-remove-nitrates-from-water.

CHAPTER 8
Perchlorate Removal

8.1 Introduction

Perchlorate is both a naturally occurring and anthropogenic chemical. Perchlorate exposure has been linked to hypothyroidism. Ingested perchlorate can interfere with iodine uptake into the thyroid gland, which can disrupt the functionality, leading to reduction of thyroid hormone production. Reduction of the thyroid hormone can lead to hormone growth defects in infants and young adults.

Sources of perchlorate exposure can be traced to both soil and groundwater contamination. This can be problematic when groundwater tables are contaminated, especially if there are traces of perchlorate within the soil in between the upper and lower groundwater tables.

This chapter describes:

- Properties and chemistry associated with the forms of perchlorate found in groundwater and surface water sources;
- Regulations associated with perchlorate, specifically, the US Environmental Protection Agency (USEPA), State of California, and World Health Organization (WHO);
- Source water quality parameters that can impact the selection of treatment alternatives suitable for perchlorate removal;
- An overview of the different treatment alternatives that can be used for perchlorate removal;
- An overview of bench and pilot testing requirements;
- Design guidelines, including calculations and residuals generation; and
- A design example.

8.2 Perchlorate Occurrence and Chemistry

8.2.1 Occurrence and Uses

Perchlorate can be found in naturally occurring environments and from industrial activity. Naturally occurring perchlorate is formed from mineral association of nitrate of soda caliche deposits which has been used in fertilizer. People and animals can also be exposed to perchlorate by ingesting food grown in perchlorate-contaminated soils. Although perchlorate can be found in naturally occurring environments, perchlorate contamination is primarily due to industrial activity. Perchlorate is mobile and extremely stable in water.

Perchlorate is commonly used in the aerospace industry as an oxidizing agent for solid propellant rockets and missiles. Other uses of perchlorate include but are not limited to fireworks, vehicle airbags, flares, matches, and explosives. In some cases, nitrate is a co-contaminant with perchlorate from industrial activity because ammonium nitrate is also frequently found in explosives and rocket fuels. Trichloroethylene is another common co-contaminant when rocket fuel is the source.

8.2.2 Chemistry

Perchlorate is an inorganic anion (ClO_4^-). In the environment, perchlorate is found as a salt with an accompanying cation. The most common form of perchlorate is ammonium perchlorate. Other forms of perchlorate include, but are not limited to sodium perchlorate, potassium perchlorate, and perchloric acid.

Chemical composition of the anion consists of four oxygen molecules and one chlorine molecules. A summary of common compounds is provided in Table 8-1.

Perchlorate is highly soluble and mobile in water which can lead to the development of large groundwater plumes. Although perchlorate is soluble in water, it is rarely found in air due to the low vapor pressure (WHO 2016).

Table 8-1 Common perchlorate compounds

Compound	Molecular Formula	CAS No.
Perchlorate anion	ClO_4^-	14797-73-0
Ammonium perchlorate	NH_4ClO_4	7790-98-9
Sodium perchlorate	$NaClO_4$	7601-89-0
Potassium perchlorate	$KClO_4$	7778-74-7
Perchloric Acid	$HClO_4$	7601-90-3

Source: WHO 2016

8.3 Perchlorate Regulations

Currently, there are no national enforceable regulations on perchlorate in drinking water and only two states have enforceable maximum contaminant levels (MCLs) for perchlorate: Massachusetts (2 µg/L) and California (6 µg/L) (USEPA 2017).

At the time this document was written, the state of California is reviewing their perchlorate MCL. The detection level for purposes of reporting (DLR) perchlorate was reduced from 4 µg/L to 2 µg/L in 2017.

Perchlorate clean-up sites are regulated under the Comprehensive Environmental Response, Compensation, and Liability ACT (CERCLA) by USEPA. However, there are instances where there is regulatory overlap with drinking water authorities and USEPA.

8.3.1 USEPA Regulations

Currently, USEPA does not have an enforceable MCL. The recommended USEPA public health advisory for perchlorate is 15 µg/L.

8.3.2 WHO Guidelines

WHO guidelines recommend total perchlorate exposure in humans to be less than 0.07 mg/L (WHO 2017).

8.3.3 State Regulations

As previously stated, only two states currently have enforceable MCLs for perchlorate in drinking water. A closer look at regulations in California are discussed to illustrate the potential of multi-agency involvement in perchlorate treatment.

In California, the Department of Toxic Substance Control (DTSC) typically requires corrective actions in closure/post-closure of permitted hazardous waste management sites. A given facility may already have remedial actions and site mitigations planned under the Federal Superfund Program or the State of California Mitigation Program. If the contamination includes surface and/or groundwater contamination, the State Resource Control Board (SWRB) or the Regional Water Quality Control Board (RWQCB) has an overlapping jurisdiction with DTSC.

California remedial plans require the following items to be evaluated before selecting a treatment alternative (ITRC 2008):

- Health and safety risks
- Beneficial uses of site sources
- Effect of the remedial actions on groundwater resources
- Site-specific characteristics
- Cost-effectiveness of alternative remedial action measures
- Potential environmental impacts of remedial actions

8.4 Perchlorate Treatment Options

Proven methods for perchlorate treatment processes include the following:

- Ion exchange (IX)
- Granular activated carbon (GAC)
- Reverse Osmosis (RO)
- Nanofiltration/ultrafiltration (NF/UF)
- Electrodialysis

Although there are a number of treatment options for perchlorate, the focus of this chapter is on using ion exchange for treatment. Since perchlorate is a negatively charged ion, it will be removed by a strong or weak base anion exchange resin. In drinking water, a perchlorate selective strong base anion (SBA) single-use resin is the most common method of removal. A weak base anion (WBA) resin would be appropriate only where there are very high concentrations of perchlorate, for example greater than 0.1 mg/L. An SBA regenerable resin has been used for drinking water but has been phased out due to the waste stream with high brine and perchlorate that needs to be disposed of.

When evaluating treatment options for perchlorate, characterization of raw water and identifying treatment objectives are critical during design development. Table 8-2 can be used to summarize common constituents important in the design of ion exchange units for perchlorate treatment.

Competing anions include but are not limited to: sulfate, nitrate, chloride, and alkalinity. Coordination with the resin supplier is recommended regarding competing ions.

Depending on the water quality, pretreatment may be necessary if the influent water has a large concentration of the following constituents: total suspended solids (TSS), oil and grease, iron and manganese, oxidants, microbes, total organic carbon/volatile organic carbon (TOC/VOC), or high propensity for scaling.

While perchlorate can be treated with either single-use or regenerable ion exchange, single-use resins have largely replaced regenerable systems for perchlorate treatment. Regenerable ion exchange produces a waste stream of brine containing perchlorate which requires environmentally conscious disposal that is not needed for single-use resins.

8.4.1 Strong-Base Anion Single-Use Ion Exchange Resin

Strong base perchlorate selective anion exchange resins for perchlorate removal are typically single use supplied usually in the chloride form. The resin generally will quickly come to equilibrium form to match the influent water chemistry within a few hundred bed volumes (BV).

Perchlorate selective resin can be either gel or macroporous, though the structure generally does not affect performance.

The selectivity for perchlorate on a perchlorate selective resin is very high.

$$ClO_4^- > NO_3^- > SO_4^{2-} > Cl^- > HCO_3^- > F^-$$

Table 8-2 Perchlorate ion exchange design backgroud water quality

Description	Units	Raw Water		
		Min	Avg	Max
Operational flow rate	gpm			
Operational schedule	hour/day			
Daily volume (average)	gallons			
Alkalinity (as $CaCO_3$)	mg/L			
Chloride	mg/L			
Sulfate (SO_4^-)	mg/L			
Nitrate (as NO_3^-)	mg/L			
Nitrate (as N)	mg/L			
Fluoride	mg/L			
Silica SiO_2	mg/L			
Carbon dioxide CO_2	mg/L			
Perchlorate	µg/L			
Arsenic	µg/L			
Hexavalent chromium	µg/L			
Uranium	µg/L			
Calcium (as $CaCO_3$)	mg/L			
Magnesium (as $CaCO_3$)	mg/L			
Sodium	mg/L			
Potassium	mg/L			
Iron, Total	mg/L			
Manganese, total	mg/L			
pH	SU			
Oxidation reduction potential (ORP)	mv			
Total dissolved solids (TDS)	mg/L			
Suspended solids	mg/L			
Oil and grease	mg/L			
Total organic carbon (TOC)	mg/L			

Uranium is also removed by the perchlorate resin, but the level of removal by different resins may vary. Uranium loading should be monitored in spent resin when profiling for disposal.

The perchlorate removal reaction is illustrated in this equation:

$$ClO_4^- + Resin\text{-}Cl^- \leftrightarrow Resin - ClO_4^- + Cl^-$$

As mentioned above, the nitrate and sulfate and other major anions will come to equilibrium within a few days so the resin does not continuously release chloride during operation.

Single-use ion exchange resins for perchlorate will last from months to years in the field dependent upon the water quality and the longevity of the resin, which makes the technology economical for end users. Single-use resin avoids the perchlorate waste streams in the regeneration process. The liability for perchlorate resin disposal may be reduced when the resin is incinerated. Disposal of spent resin should be considered in the operational costs as discussed in Section 8.7, Disposal Options.

8.5 Bench and Pilot Testing

Chapter 13 of this handbook discusses the general process of piloting an ion exchange system. This section focuses on the specific considerations for perchlorate removal. As with all bench and pilot testing, the test should mimic and be scalable to the anticipated full-scale system. Based on resin manufacturing modeling, a pilot study may be beneficial to validate full-scale design parameters including surface loading rate, throughput, empty bed contact time and waste characteristics. The pilot column testing generally includes multiple columns that allow the comparison of different resin types. Piloting can also help determine if pretreatment is necessary.

A comprehensive water quality analysis should be performed that includes all the anions and cations. Chapter 2 presents a table describing the typical water quality analyses. Also see Table 8-2.

In many cases resin suppliers have enough experience with perchlorate selective resins such that pilot testing may not be needed. Suppliers will provide modeling to determine throughput based on the water chemistry.

8.6 Design Guidelines

Perchlorate selective ion exchange resins for drinking water applications are single-use. A well-designed treatment system not only implements best practices for resin system sizing and equipment configuration, but also provides pretreatment as needed and considers operation and maintenance aspects in the overall design, including ancillary equipment like feed pumps.

Since perchlorate resins are single-use, they are not typically backwashed and therefore pretreatment is especially important. Pretreatment for suspended solids removal is particularly important for single-use resins. Solids buildup on the resin bed can lead to uneven flow distribution and loading within the bed due to blockage and channeling. Typically, suspended solids removal is achieved by providing prefiltration using bag filters or cartridge filters.

Iron and manganese are common co-contaminants in perchlorate removal applications, particularly in groundwater sources. While low levels of iron and manganese that remain in solution can pass through the resin system without a significant impact, pretreatment should be considered, particularly if it is determined that particulate iron or manganese are present (Purolite n.d.).

Refer to Chapter 2: Fundamentals of Ion Exchange, Section 2.3, for a more detailed description of water quality pretreatment goals and options.

8.6.1 Resin Vessel Design Guidelines

Below are the critical elements for vessel design to treat perchlorate using for a six-vessel resin in lead/lag operation:

- Empty bed contact time (EBCT): 1.5–3 min per vessel
- Hydraulic loading rate: 6–18 gpm/ft^2 (15–45 m/h)
- Resin depth: 3 ft minimum at loading rates up to 12 gpm/ft^2; 3.7 ft at loading rates greater than 12 gpm/ft^2
- Specific flow rate: 1–5 gpm/ft^3 (8–40 BV/h)
- Configuration: lead/lag vessel (series) configuration
- Minimum pretreatment: solids removal to 5–10 microns

The kinetics of perchlorate resin are relatively fast as shown by the EBCT. Resin vessels must be properly designed to handle the faster hydraulics indicated above.

8.7 Disposal Options

The perchlorate selective single-use ion exchange resin system has two potential sources of wastewater: backwash waste and rinsing waste primarily associated with resin system startup and resin changeouts.

Backwashing: The purpose of initial resin backwashing is typically for the removal of resin fines that may be present in the delivered resin and grading of the resin bed as much as possible. The recommendations for initial backwashing will vary depending on the resin supplier and the resin's physical properties. Some suppliers recommend backwashing for 30–90 min at as much as 50 percent resin bed expansion, while others take the position that backwashing is not necessary. It is good practice to include provisions in the design for backwashing the resin. Even if the resin that is initially installed does not require backwashing, this may change if in the future if a different resin is used.

Recommended backwash rates will vary depending on the water temperature and a given resin's properties, but typical backwash rates to achieve 50 percent bed expansion at 55°F are approximately 1–2 gpm/ft^2. The water source for resin backwashing should be considered in the design. If chlorinated water sources (e.g., finished water) are proposed, the impact of the short-term oxidant exposure on the resin should be reviewed with the resin supplier. If direct sewer discharge at the desired backwash rate is possible, backwash waste handling may be straightforward. If direct discharge is not possible or if the sewer is not available, provisions for temporary or permanent waste handling infrastructure should be provided.

In general, once placed into service, backwashing of the bed is not recommended, except as a last resort to try to alleviate a resin solids buildup or clogging issue that would otherwise require premature changeout of a resin bed. Once resin has been placed into service, it will begin to develop zones in the bed with the top of the bed characterized by spent resin, the bottom of the bed characterized by "fresh" resin, and the area in between acting as a mass transfer zone for the target contaminant, which slowly descends through the bed as the resin capacity is exhausted. Backwashing of the bed once in service will disturb the mass transfer zone and potentially lead to premature contaminant breakthrough.

Rinsing: Perchlorate selective ion exchange resins typically need an initial forward flow rinse prior to use in potable water applications. Most resins have an initial minimum rinsing requirement that can sometimes be performed prior to delivery. Where rinsing will not be performed prior to delivery, a plan for temporary or permanent infrastructure for waste handling will need to be considered where direct discharge to a sewer is not possible. A resin application where the resin is not pre-rinsed prior to delivery will typically require a minimum rinse of 20 bed volume (BV). The required minimum BV for initial rinsing should be provided by the resin supplier.

Perchlorate selective ion exchange resins delivered in an unbuffered form will temporarily impact water quality before the resin reaches equilibrium with the background water chemistry. Resin behavior in an application will depend on the type of resin and the background water chemistry.

8.7.1 Resin Disposal Options

The disposal of spent perchlorate resin is subject to federal, state, local, and tribal regulations. Depending on the amount of perchlorate in the resin, disposal options may be limited depending on local or state laws. For single-use resins, disposal via high temperature incineration allows for destruction of the perchlorate and cradle-to-grave responsibility. Resins can be also be landfilled depending upon the local regulations.

Generally, the resin will need to be profiled before disposal subject to the waste acceptance criteria of the waste facility. Testing may include, toxicity characteristic leaching procedure (TCLP), CAM Metals 17, uranium loading, and fish bioassay. Typically, the spent resin is non-hazardous.

On some perchlorate selective resins depending on functionality, there is a potential for uranium loading; although this not common. If this occurs, operators need to monitor uranium levels and determine if uranium accumulation will trigger uranium disposal and handling regulation requirements. Personnel and facilities licensed to handle and dispose uranium may be needed for these types of resins. See Chapter 5, Radionuclide Removal, for additional information.

8.8 Design Example

A throughput projection can be done by contacting the resin supplier.

8.8.1 SBA Single-Use Example

A lead/lag vessel configuration is selected. In this example, the raw water perchlorate is 13 ppb and the goal is to achieve less than 4 ppb out of the lead vessel, and less than 1 ppb out of the lag vessel. For perchlorate, 100 percent of the stream is treated, there is no blending. A summary of raw water quality is provided in Table 8-3. A design example for groundwater perchlorate treatment of an ion exchange unit in a lead/lag configuration is provided for reference (see Table 8-4).

The following steps are recommended when designing a perchlorate IX treatment system:

- Determine available equipment diameter:
  - 12 ft diameter vessels: 2 × radius: r = 6 ft
  - Determine vessel cross sectional area: = 113 ft^2
  - Determine linear velocity
    - Flow rate/vessel cross sectional area = 1,200 gpm/113 ft^2 = 10.6 gpm/ft^2
    - This falls into the acceptable range of 6–18 gpm/ft^2 (15–45 m/h)
- Determine volume of resin to achieve the minimum bed depth by varying the number of supersacks. With a linear velocity of 10.6 gpm/ft^2, we need a minimum bed depth of 3 ft.
  3 ft × 113 ft^2 = 339 ft^3
  - It is much easier to load and unload based on using full supersack volumes.
  - 42 ft^3 per supersack × 9 supersacks = 378 ft^3
  - In this example, we decided to add an additional supersack for prolonged bed life.
  - 42 ft^3 per supersack × 10 supersacks = 420 ft^3
- Determine specific flow rate:
  Specific flow rate = flow (gpm)/resin volume (ft^3)
  Specific flow rate = 1,200 gpm/420 ft^3 = 2.9 gpm/ft^3

Table 8-3 Raw groundwater parameters

Raw Water Parameter	Value
Flow rate	1,200 gpm
Perchlorate	13 μg/L
Alkalinity as $CaCO_3$	80 mg/L
Chloride	170 mg/L
Sulfate	30 mg/L
Nitrate as N	19 mg/L
pH	7.5

Table 8-4 SBA example design parameters

Parameter	Value
Flow rate	1,200 gpm
Number of trains	1
Number of vessels per train	2 Lead / lag
Vessel diameter	12 ft
Volume of resin in supersack	42 ft^3
Number of supersacks	10
Media volume per vessel	420 ft^3
Media volume total onsite	840 ft^3
Vessel area	113 ft^2
Bed depth: min 3 ft	3.7 ft
Linear velocity: goal 8–12 gpm/ft^2	10.6 gpm/ft^2
Specific flow rate: goal 1–5 gpm/ft^3	2.9 gpm/ft^3
EBCT	2.6 minutes
BV/hour	22.9 BV/h
Estimated throughput—determined by modeling from the resin supplier	86,000 BV
Days between exchanges	156 days

BV: bed volume, EBCT: empty bed contact time

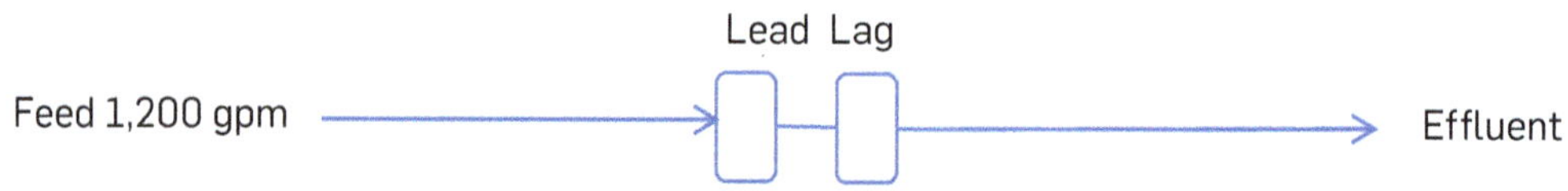

Figure 8-1 Perchlorate single-use design

Courtesy of Purolite Corporation

- Determine EBCT:

  EBCT = vessel volume/treated flow rate

  EBCT = 420 ft^3 × 7.48 gal/ft^3/1,200 gpm = 2.6 min
- Calculate the days between exchanges:

  Estimated throughput as provided by resin manufacturer = 86,000 BV

  86,000 BV / 22.9 BV/h = 3755.5 h = 156 days

Figure 8-1 shows the ion exchange system in a lead/lag configuration.

REFERENCES

ITRC. 2008. "Remediation Technologies for Perchlorate Contamination in Water and Soil," Technical/Regulatory Guidance, March. http://www.eosremediation.com/download/Perchlorate/ITRC%20PERC-2.pdf.

Purolite. n.d. "Design and Operational Guidelines: PFAS-Selective Single Use Ion Exchange Resin for Drinking Water Systems." Application Guide. Design and Operational Guidelines. https://www.purolite.com/dam/jcr:e7dc9abb-abeb-4363-b46a-e44f1ae4ce18/PFAS-Selective%20Single%20Use%20Ion%20Exchange%20Resin%20for%20Drinking%20Water%20Systems_Application%20Guide.pdf.

USEPA (US Environmental Protection Agency). 2017. Technical Fact Sheet – Perchlorate. USEPA. 15 September 2017. https://www.epa.gov/sites/production/files/2017-10/documents/perchlorate_factsheet_9-15-17_508.pdf

WHO (World Health Organization). 2016. Perchlorate in Drinking Water. WHO. 2016. https://www.who.int/water_sanitation_health/water-quality/guidelines/chemicals/perchlorate-background-jan17.pdf?ua=1

CHAPTER 9
Chromium Removal

9.1 Introduction

Chromium found in drinking water can come from natural or manmade sources. One of the most famous cases of drinking water chromium contamination occurred in Hinkley, Calif., as depicted in the movie *Erin Brockovich*. The Hinkley contamination was due to Chromium 6—one of the nine oxidation states of chromium.

This chapter describes how ion exchange in both non-regenerable and regenerable forms can be used for chromium removal from drinking water. This chapter describes:

- properties and chemistry associated with the forms of chromium found in groundwater and surface water sources;
- regulations regarding chromium: specifically the US Environmental Protection Agency (USEPA), the World Health Organization (WHO), and California state-specific regulation;
- source water quality parameters that can impact the selection of ion exchange resins suitable for chromium removal;
- an overview of the types of resins that can be used for chromium removal;
- an overview of bench and pilot testing requirements;
- design guidelines including calculations and residuals generation; and
- a design example.

9.2 Chromium Occurrence and Chemistry

9.2.1 Occurrence and Uses

Chromium is an odorless and tasteless metallic element and is one of the most abundant metals in the earth's crust. It can be found in rocks, plants, soil, and animals (USEPA 2015). Chromium does not exist in nature by itself and instead is found as a compound in many different chemical configurations. Naturally occurring chromium can enter our waters from the interaction of groundwater with the soil matrix ("Chromium | History, Uses, Facts, Physical & Chemical Characteristics" 2018).

Manmade sources of chromium can also enter our ground and surface waters. Understanding the manmade uses of chromium is important as these uses can assist in the understanding of the potential sources of the contamination. The uses include:

- production of metal alloys such as stainless steel as chromium can improve corrosion resistance,
- a coating agent for various surfaces,
- the manufacture of dyes – the word chromium derives from the Greek word chroma which means color,
- leather manufacturing,
- cleaning products, and
- corrosion control for open and closed cooling water systems.

9.2.2 Chemistry

Chromium can be found in one of nine valence (or oxidation) states ranging from -2 to +6 and is measured as total chromium. In drinking water, it is most common in the +6 state (known as hexavalent chromium). Hexavalent chromium can also be expressed as CrO_4^{2-}, Cr(VI), hex chrome, chromate, chrome VI, or as a dichromate ion ($Cr_2O_7^{2-}$).

Chromium can also be found in the +3 valence state (trivalent chromium) or Cr(III). Trivalent chromium is necessary for human nutrition and according to the Institute for Medicine Food and Nutrition Board 25-45 µg Cr(III) daily is necessary (Edzwald 2011). Trivalent chromium is the most stable state of chromium; however, hexavalent chromium creates adverse health effects including cancer.

In most drinking water sources chromium will exist as hexavalent chromium, though it is good practice to analyze for both total chromium and hexavalent chromium.

9.3 Chromium Regulations

9.3.1 Regulations and Standards

In 1991 chromium was regulated with an enforceable maximum contaminant level (MCL) of 0.1 mg/L (100 µg/L) in the US by the EPA. The drinking water standard is for total chromium (i.e., all forms of chromium). The current standard is based on potential adverse dermatological effects from long-term exposure, such as allergic dermatitis (skin reactions) and not on newer research showing that hexavalent chromium can cause cancer. EPA regularly re-evaluates drinking water standards and, based on new science on hexavalent chromium, began a rigorous and comprehensive review of its health effects in 2010. Although EPA has been reviewing the chromium standard for some years, there has been no change to the regulations. California has the most stringent standard for chromium with a MCL of 50 µg/L for total chromium.

In 2014 the state of California was the first State to release a MCL specific to hexavalent chromium of 0.01 mg/L (10 µg/L). However, in 2017 the California Manufacturers and Technology Association sued the California State Water Resources Control Board over the regulation and the court ruled that the standard was invalid, stating the State Water Board "did not adequately document why the MCL was economically feasible."

At this time, the state of California is attempting to expedite the process of setting a new MCL for hexavalent chromium. Municipalities with treatment systems already installed and operating are encouraged by the state to continue to operate these facilities. The MCL of 50 µg/L for total chromium remains in place while a new MCL is being determined.

9.3.2 International Regulations and Guidelines

WHO in its 1996 Guidelines for Drinking Water Quality, 2nd ed. provides a guideline of 0.05 mg/L (50 µg/L) for total chromium (Organization 1993).

In 2018, Health Canada set a Maximum Acceptable Concentration (MAC) of 0.05 mg/L (50 µg/L) total chromium. According to Health Canada, this MAC protects against both cancer and non-cancer effects from exposure of hexavalent chromium (Canada 2018).

9.4 Ion Exchange Resin Treatment Options

The EPA lists coagulation/filtration, reverse osmosis, lime softening, and ion exchange as best available technologies (BATs) for chromium removal (USEPA 2007). This section focuses on treatment options using ion exchange resin technology. As discussed earlier, chromium can exist in several oxidation states and the oxidation states of the chromium species should be analyzed when the best treatment technology is being considered.

Weak base anion (WBA) and strong base anion (SBA) exchange resins can remove over 95 percent of hexavalent chromium concentrations typically present in drinking water (Canada 2015). Chromium-3 is positively charged and would require cation exchange, however this is rarely found in drinking water. As with all water treatment applications, a detailed raw water analysis is needed for proper resin selection and system design. Water parameters needed include: pH, TDS, TSS, alkalinity, nitrate, sulfate, chloride, uranium, hardness, and temperature.

WBA resin is typically used on a one-time (single-use) basis whereas SBA resin is typically used on a regenerable basis. In some limited cases, SBA resin can be single-use, especially when treating very low TDS waters. With WBA resin, it is necessary to reduce the influent water pH to approximately 5.5–6.0 using an acid. After treatment the pH will need to be raised again, typically with caustic, before sending to the distribution system. Once the resin is spent, it is then either disposed of as a hazardous waste or sent offsite to a chromium reclamation facility.

With SBA resin, no pH adjustment is normally needed, and the resin can be reused multiple times by using a brine solution periodically to elute the chromium from the resin. The SBA design must include proper treatment and disposal of the spent chromium-containing brine. Choosing between WBA and SBA resin technology will depend on:

a. the impact of competing anions (e.g. SO_4^{2-}, NO_3^-, HCO_3^-, Cl^-) on resin capacity,
b. the size of the treatment plant and the availability of operators for daily maintenance,
c. the target treatment level for hexavalent chromium,
d. whether acid and caustic for pH control with WBA resin can be accommodated, and
e. the site options available for handling and disposal of chromium-containing brine.

Each treatment site will have its combination of factors that must be carefully evaluated and modeled. As a rule of thumb, WBA resin tends to be favored for water with increasing levels of sulfate and total dissolved solids. On the other hand, high sulfate levels usually mean more frequent regeneration of SBA resins and therefore increasing cost for regenerant and wastewater disposal. Below is a more in-depth discussion of the merits of both types of resins.

Ion exchange is very effective at removing chromium, but not 100 percent effective. The amount of chromium that is not removed is called leakage and can vary based type of resin used and other constituents in the water. Typical leakage can be expected to be around 1–2 µg/L.

9.4.1 Weak Base Anion (WBA) Exchange Resins

A WBA resin must be protonated (addition of a proton) before it can remove hexavalent chromium and other anions from the water. Reducing the pH of the influent water contacting the resin to a slightly acidic pH range of 5.5–6.0 activates (or protonates) the resin. At the same time, reducing the pH converts most of the hexavalent chromium from its divalent to its monovalent state—this essentially doubles the loading capacity of the resin for hexavalent chromium and facilitates single-use operation of the resin. Consideration should be given to the following design and operating factors:

- The pH of the influent water must be reduced to less than 6 before the WBA resin.
- The bicarbonate alkalinity of the raw water will determine the acid dosage needed. The choice of acid for pH adjustment depends in part on the water chemistry, particularly the calcium level, and the requirement for NSF certification. HCl use is not dependent on calcium level, so can be utilized for nearly all applications. If using H_2SO_4, and the calcium level is on the higher side (>25 percent of all cations), there is the chance for $CaSO_4$ precipitation, so it should not be used.
- The pH of the WBA treated water must be re-adjusted upward to the influent levels before the water is sent to the distribution system.
- Most drinking water well quality is amenable to WBA treatment.
- Any uranium present in the influent water will also be removed by the resin; this can sometimes limit loading capacity and can determine the disposal cost for the spent resin.
- WBA resin can generally treat hundreds of thousands of bed volumes (BV) of water before it is spent, making it economically feasible in many cases.
  - Epoxy-amine and phenol-formaldehyde type polymer group WBA resins are often used for this function.
- If using a phenol-formaldehyde type WBA resin, preconditioning would need to be coordinated with the manufacture generally needed to prevent the release of low levels of formaldehyde into the treated water.

9.4.2 Strong Base Anion Exchange Resins

SBA exchange resins are effective over a wider pH range and do not typically require pH adjustment. Water passes through the resin where hexavalent chromium and other anions are exchanged for chloride initially.

SBA resin is more selective for hexavalent chromium versus other major anions typically present:

$$CrO_4^{2-} > SO_4^{2-} > NO_3^- > Cl^- > HCO_3^- > F^-$$

Depending on the water chemistry – specifically sulfates – SBA resins can be used as either a single-use or a regenerable resin.

The reaction is:

$$CrO_4^{2-} + Resin - Cl_2 \leftrightarrow R\text{-}CrO_4 + 2Cl^-$$

SBA resins tend to have a much higher affinity for hexavalent chromium versus other major anions, allowing loading capacities typically up to 20,000 BV and even higher for low TDS waters (<10 mg/L). This reaction is reversible if a large excess of chloride is used, such as during the regeneration step.

9.4.2.1 Single-Use SBA Resin

When sulfates are very low (<10 mg/L) high capacity SBA resins can last months to years in the field. Economic evaluation is necessary to determine the throughput and change-out costs as well as the capital savings of a single-use system. The advantages of a single-use SBA resin are no generation of liquid waste (i.e., no spent brine), and no need for pH adjustment.

9.4.2.2 Regenerable SBA Resin

Using a regenerable SBA resin is feasible when the concentrations of sulfate and other competing anions are moderate and regeneration is needed every few weeks or longer. If the SBA resin needs to be regenerated more often, then the use of a WBA resin may be preferable.

When a vessel of resin has reached its change-out criteria, the vessel is taken out of service in preparation for regeneration. During the regeneration step, several bed volumes of sodium chloride (NaCl) brine regenerant are pumped through the resin at a high enough concentration to strip the hexavalent chromium and other anions from the resin and put the resin back in the chloride form. The brine will contain high concentrations of both sodium chloride and hexavalent chromium. The system then goes through a slow rinse to displace the last contaminants followed by a fast rinse, mainly to displace the NaCl. The vessel is then ready to be placed back in service.

High hexavalent chromium in the spent brine may result in it being classified as hazardous. Typically, a brine analysis is needed prior to disposal.

9.4.2.3 Brine Disposal Options

In most cases, spent brine cannot be discharged to a sanitary sewer system due to the impact on the wastewater treatment plant. The impacts on wastewater treatment plant discharges and National Pollutant Discharge Elimination System (NPDES) requirements (USEPA 2007) must be evaluated for each system.

Brine reuse has been studied by determining where the chromate concentrations are highest in the brine stream with an elution study (usually between 2 and 4 BV). The high chromate portion of the brine waste can be disposed of and the rest of the spent brine can be reused. This practice reduces the volume of waste that needs to be hauled off (Jacobs 2019). Disposal of the brine waste would generally consist of hauling the liquid waste to a hazardous waste facility.

Spent brine will also concentrate other anions in the water. Complete characterization is necessary to ensure the uranium concentration as well as the other ionic species are compatible with the requirements of the disposal site.

9.5 Pilot Testing

Chapter 13 of this handbook discusses the general process of piloting an ion exchange system. This section focuses on the specific considerations for chromium removal. As with all bench and pilot testing, the test should mimic and be scalable to the anticipated full-scale system. Bench-scale testing can help determine the optimal empty bed contact time (EBCT) to conduct subsequent experiments, screen resins to identify best performing options, and demonstrate treatment performance through multiple loading and regeneration cycles.

A comprehensive water quality analysis should be performed that includes all of the anions and cations. Chapter 2 presents a table describing the typical water quality analyses. With respect to chromium, both total chromium and chromium 6 should be analyzed. Furthermore, uranium, sulfate, pH, alkalinity and nitrate all will affect the selection of the most appropriate anion exchange resin and sampling should be performed to enable an assessment of the variability of these parameters.

Based on manufacturer modeling, a pilot study may be beneficial to provide full-scale design parameters including surface loading rate, throughput, EBCT and waste characterization. The pilot column tests generally have multiple trains that allow the comparison of different resin types. Piloting can also help determine hydrochloric acid and caustic dosages that may be needed if WBA resin is used.

9.6 Design Guidelines

Recommended design guidelines for regenerable resins are further described below.

Pretreatment:

- Prefilters to prevent total suspended solids (TSS) accumulation in the resin bed is vital to satisfactory operation. The resin will act as an absolute filter producing water with an effluent turbidity of less than 0.3 NTU. The longer the resin will be in service, especially with single-use ion exchange resins, the more important proper design and operation of a prefiltration system is to the resin performance. A particle size distribution analysis can help determine the size of incoming solids. If the well produces any sand, a de-sander is also recommended. The feed turbidity to the system should be less than 1.0 NTU.
- Resin performs best when foulants are eliminated. If any of these are present in the water in concentrations high enough to affect resin performance, they need to be adequately removed:
  - Iron and manganese > 1 ppm
  - Oil and grease – any concentration is detrimental
  - Oxidants – any concentration is detrimental

- Langelier Saturation Index – ensure there is no scaling potential, < 0
- TOC > 2 ppm

The recommended design guidelines for chromium removal are:

WBA design parameters:

- Resin: WBA
- Resin depth: 3 ft minimum (0.91 m)
- Hydraulic loading Rate: 8–12 gpm/ft² (20–30 m/h)
- Specific flow rate: 1–5 gpm/ft³ (8–40 BV/h)
- EBCT: 2–3 min minimum
- pH must be lowered to between 5.5 and 5.9. HCl, H_2SO_4 liquid acids can be used, or CO_2 can be injected as a gas
    - After ion exchange treatment, additional pH adjustment may be required. Chemicals such as sodium hydroxide and lime can be used to meet the finished water quality requirements.
- Vessels are generally designed for a lead/lag configuration.

SBA design parameters:

- Resin: SBA
- Resin depth: 3 ft minimum (0.91 m)
- Hydraulic loading rate: 8–12 gpm/ft² (20–30 m/h)
- Specific flow rate: 1–5 gpm/ft³ (8–40 BV/h)
- EBCT: 2–3 min minimum

If regenerating SBA resin:

- Regenerant solution: Sodium chloride brine
- Regenerant concentration: 10–12 percent NaCl
- Regenerant dosage: approx. 35 lb NaCl/ft³ of resin

Regeneration steps:

- Backwash (as needed): Use manufacturer's backwash curves to achieve 50 percent bed expansion for anywhere from 1–3 BVs depending on solids loading
- Brine step with regenerant solution: 5 BVs with 12 percent NaCl, about 35 lb/ft³ of NaCl
- Slow rinse with treated water: Typically 0.25–0.5 gpm/ft³ (2–4 BV/h) for 30–40 min
- Fast rinse with treated water: Typically, at the service flow rate for 10–20 min.

A key component in the design of ion exchange systems for chromium removal is the estimation of the amount of water that can be treated before the resin is exhausted (referred to as throughput). The throughput is based on the water quality parameters and is generally obtained by the following:

- evaluation of similar systems,
- from the resin suppliers each of which have models that allow an estimation of throughput, and
- from pilot testing.

9.7 Design Examples

Design examples have been provided below for: WBA, SBA single-use, and SBA regenerable ion exchange systems for chromium removal.

9.7.1 WBA Example

In this example, the groundwater source has an influent hexavalent chromium of 20 µg/L. The goal is to treat water to less than or equal to 8 µg/L. Therefore, partial treatment and blending is most economical. The sulfate level is 120 mg/L, so SBA regeneration is not economical due to frequency of regeneration. WBA is the best solution in this case because of the high sulfate. Table 9-1 summarizes the raw water quality.

A throughput projection can be done by contacting the resin supplier. In the example below a pilot test was performed and 145,000 BV was achieved. From experience, the WBA resin has an estimated Cr(VI) leakage of 1 µg/L.

Because this is a regenerable system, the raw water uranium level will not impact the hexavalent chromium capacity. However, depending on the length of the run and therefore the accumulation of uranium, disposal of the regenerant waste may be a consideration. Before the design is completed and method of disposal selected, a calculation of regeneration frequency and the corresponding uranium inventory prior to that regeneration should be made. As discussed in Chapter 5, disposal of uranium residuals is state dependent and may be allowed without scavenging the uranium from the waste. If the uranium level in the

Table 9-1 WBA example raw groundwater parameters

Raw Water Parameter	Value
Well pump flow rate	950 gpm
Total chromium	20.4 µg/L
Chrome VI	20 µg/L
Alkalinity	120 mg/L as $CaCO_3$
Chloride	120 mg/L
Sulfate	120 mg/L
Nitrate	5 mg/L as N
Uranium	2 µg/L
pH	7.5

regenerant waste exceeds the level of the receiving publicly owned treatment works (POTW), the uranium can be scavenged from the waste using a single-use system. Spent resin from that single-use system would then be considered technologically enhanced naturally occurring radioactive materials (TENORM), with licensed disposal accordingly. Refer to Chapter 5: Radionuclide Removal for additional information.

- Determine treated flow:
  - Bypass percentage = Effluent concentration goal – Leakage concentration) / Influent concentration – Leakage concentration

$$\frac{8\text{ ppb} - 1\text{ ppb}}{20\text{ ppb} - 1\text{ ppb}} = 36.8\%$$

  - Bypass flow = total flow rate x bypass percentage

$$950\text{ gpm} \times 36.8\% = 350\text{ gpm}$$

  - Treated Flow = total flow – bypass flow

$$950\text{ gpm} - 350\text{ gpm} = 600\text{ gpm}$$

Design Steps:

- Determine available equipment diameter using the linear velocity criteria:
  - Assume linear velocity: 12 gpm/ft²
  - Vessel area = treated flow / linear velocity

$$\frac{600\text{ gpm}}{12\text{ gpm/ft}^2} = 50\text{ ft}^2$$

  - Vessel diameter: $A = \pi r^2 \rightarrow d = 2\sqrt{\frac{A}{\pi}}$

$$50\text{ ft}^2 = \pi r^2 \quad d = 2\sqrt{\frac{50\text{ ft}^2}{\pi}} = 8\text{ ft}$$

- Determine media volume per vessel based on specific volume criteria:
  - Assume specific flow rate: 2.9 gpm/ft³
  - Vessel volume = treated flow / specific flow rate

$$\frac{600\text{ gpm}}{2.9\text{ gpm/ft}^3} = 207\text{ ft}^3$$

- Determine volume of resin to achieve the minimum bed depth by varying the number of supersacks:
  - Assume resin is shipped in 42 ft^3 supersacks (ss)

$$\frac{207\ \text{ft}^3}{42\ \text{ft}^3/\text{ss}} = 5\ \text{supersacks/vessel}$$

- Determine depth of resin in vessel:
  - Vessel volume / vessel area = depth

$$\frac{207\ \text{ft}^3}{50\ \text{ft}^2} = 4.14\ \text{ft} \approx 4.1\ \text{ft}$$

- EBCT:
  - EBCT = vessel volume / treated flow rate

$$207\ \text{ft}^3 \times \frac{7.48\ \text{gal}}{\text{ft}^3} \times \frac{\text{min}}{600\ \text{gal}} = 2.6\ \text{min}$$

  - 2.6 min is within the EBCT goal criteria.
- Determine number of trains to meet appropriate design conditions

Table 9-2 presents a summary of the design criteria for this regenerable WBA system. Figure 9-1 shows the basic configuration for this regenerable WBA with by-pass configuration.

Table 9-2 WBA example design parameters

Design Parameters	Value
Total flow rate	950 gpm
Influent chrome VI	20 µg/L
Leakage chrome VI	1 µg/L
Effluent concentration goal of chrome VI (blended)	8 µg/L
Bypass flow percent	36.8%
Bypassed flow rate	350 gpm
Treated flow rate	600 gpm
Number of trains	1
Number of vessels per train	2
Vessel diameter	8 ft
Flow per vessel	600 gpm
Volume per supersack	42 ft^3
Number of supersacks per vessel	5

(continued)

Table 9-2 WBA example design parameters (*continued*)

Design Parameters	Value
Media volume per vessel	210 ft^3
Media volume total onsite	420 ft^3
Vessel area	50 ft^2
Bed depth	4.2 ft
Linear velocity: goal 8–12 gpm/ft^2	11.9 gpm/ft^2
Specific flow rate: 1–5 gpm/ft^3 target per vessel	2.9 gpm/ft^3
EBCT: goal 2.5–3 min	2.62 minutes
Projected throughput	145,000 BVs
Days between exchanges	264 days

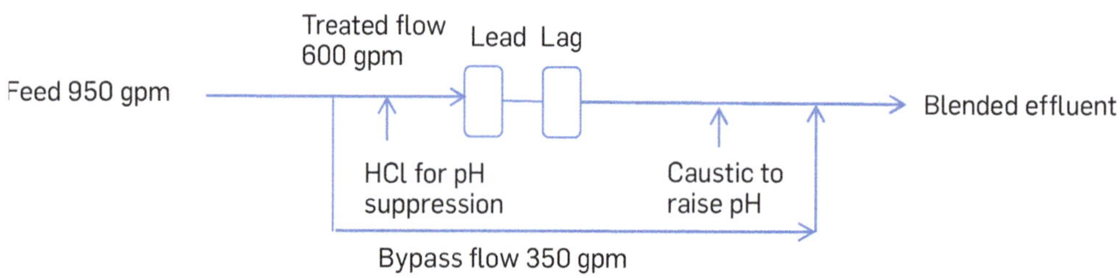

Figure 9-1 WBA design
Courtesy of Purolite Corporation

9.7.2 SBA Single-Use Example

In this example, the influent hexavalent chromium is 20 µg/L. The goal is to treat water to less than or equal to 8 µg/L. Therefore, partial treatment and blending is most economical. The sulfate level is 2 mg/L, so SBA single-use is a good option because it does not require pH adjustment. Table 9-3 summarizes raw water quality parameters.

A throughput projection can be done by contacting the resin supplier. In the example below a pilot test was performed and 100,000 BV was achieved. From experience, the SBA resin has an estimated hexavalent chromium leakage of 1 µg/L.

- Determine treated flow:
  - Bypass percentage = Effluent concentration goal – Leakage concentration) / Influent concentration – Leakage concentration

$$\frac{8 \text{ ppb} - 1 \text{ ppb}}{20 \text{ ppb} - 1 \text{ ppb}} = 36.8\%$$

Table 9-3 SBA example raw groundwater parameters

Raw Water Parameter	Value
Well pump flow rate	950 gpm
Total chromium	20.4 µg/L
Hexavalent chromium	20 µg/L
Alkalinity	120 mg/L as $CaCO_3$
Chloride	120 mg/L
Sulfate	2 mg/L
Nitrate	5 mg/L as N
Uranium	2 µg/L
pH	7.5

- Bypass flow = total flow rate x bypass percentage

$$950 \text{ gpm} \times 36.8\% = 350 \text{ gpm}$$

- Treated flow = total flow – bypass flow

$$950 \text{ gpm} - 350 \text{ gpm} = 600 \text{ gpm}$$

Design Steps:

- Determine available equipment diameter using the linear velocity criteria:
  - Assume linear velocity: 12 gpm/ft^2
  - Vessel area = treated flow / linear velocity

$$\frac{600 \text{ gpm}}{12 \text{ gpm/ft}^2} = 50 \text{ ft}^2$$

  - Vessel diameter: $A = \pi r^2 \rightarrow d = 2\sqrt{\frac{A}{\pi}}$

$$50 \text{ ft}^2 = \pi r^2 \quad d = 2\sqrt{\frac{50 \text{ ft}^2}{\pi}} = 8 \text{ ft}$$

- Determine media volume per vessel based on specific volume criteria:
  - Assume specific flow rate: 2.9 gpm/ft^3
  - Vessel volume = treated flow / specific flow rate

$$\frac{600 \text{ gpm}}{2.9 \text{ gpm/ft}^3} = 207 \text{ ft}^3$$

- Determine volume of resin to achieve the minimum bed depth by varying the number of supersacks:
  - Assume resin is shipped in 42 ft^3 superacks (*ss*)

$$\frac{207 \text{ ft}^3}{42 \text{ ft}^3/\text{ss}} = 5 \text{ supersacks/vessel}$$

- Determine depth of resin in vessel:
  - Vessel volume / vessel area = depth

$$\frac{207 \text{ ft}^3}{50 \text{ ft}^2} = 4.14 \text{ ft} \approx 4.1 \text{ ft}$$

- EBCT:
  - EBCT = vessel volume / treated flow rate

$$207 \text{ ft}^3 \times \frac{7.48 \text{ gal}}{\text{ft}^3} \times \frac{\text{min}}{600 \text{ gal}} = 2.6 \text{ min}$$

  - 2.6 min is within the EBCT goal criteria.
  - It is much easier to load and unload based on using full supersack volumes.
- Determine number of trains to meet appropriate design conditions.

Figure 9-2 shows a flow diagram for a SBA single-use resin in a lead-lag configuration with bypass flow.

Table 9-4 presents a summary of the design criteria for the SBA single use resin system.

Residuals Handling:

- Test for uranium accumulation in spent resin to determine proper disposal method.

9.7.3 SBA Regenerable Example

In this example, the influent hexavalent chromium is 20 µg/L. The goal is to treat the water to less than or equal to 8 µg/L. A partial treatment and blending is most

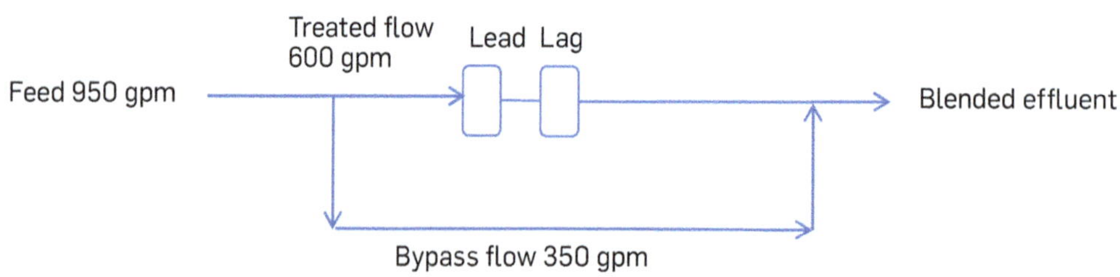

Figure 9-2 SBA single-use design
Courtesy of Purolite Corporation

Table 9-4 SBA design parameters

Design Parameters	Value
Total flow rate	950 gpm
Influent chrome VI	20 µg/L
Leakage chrome VI	1 µg/L
Effluent concentration goal (blended)	8 µg/L
Bypass flow percent	36.8%
Bypassed flow rate	350 gpm
Treated flow rate	600 gpm
Number of trains	1
Number of vessels per train	2
Vessel diameter	8 ft
Flow per vessel	600 gpm
Volume per supersack	42 ft^3
Number of supersacks per vessel	5
Media volume per vessel	210 ft^3
Media volume total onsite	420 ft^3
Vessel area	50 ft^2
Bed depth	4.2 ft
Linear velocity rate: goal 8–12 gpm/ft^2:	11.9 gpm/ft^2
Specific flow rate: 1–5 gpm/ft^3 target per vessel	2.9 gpm/ft^3
EBCT: goal 2–3 min	2.62 minutes
Projected throughput	100,000 BV
Days between exchanges	182 days

economical. The sulfate level is 25 mg/L, so SBA regenerable is a good option because no pH adjustment is needed and resin change out would be too frequent to be economical. Table 9-5 summarizes the raw water quality.

A throughput projection can be done by contacting your resin supplier. In the example below a pilot test was performed and 9,925 BV was achieved. From experience, the regenerable SBA resin has an estimated hexavalent chromium leakage of 2 µg/L.

- Determine treated flow:
  - Bypass percentage = Effluent concentration goal – Leakage concentration) / Influent concentration – Leakage concentration

$$\frac{8\text{ ppb} - 2\text{ ppb}}{20\text{ ppb} - 2\text{ ppb}} = 33.3\%$$

  - Bypass flow = total flow rate × bypass percentage

$$950\text{ gpm} \times 33.3\% = 316.7\text{ gpm}$$

Table 9-5 SBA example raw groundwater parameters

Raw Water Parameter	Value
Well pump flow rate	950 gpm
Total chromium	20.4 µg/L
Hexavalent chromium	20 µg/L
Alkalinity	120 mg/L as $CaCO_3$
Chloride	120 mg/L
Sulfate	25 mg/L
Nitrate	5 ppm as N
Uranium	2 µg/L
pH	7.5

- Treated flow = total flow – bypass flow

$$950 \text{ gpm} - 317 \text{ gpm} = 633 \text{ gpm}$$

For regenerable resin systems the vessels will be operated in parallel. Use two duty vessels and one in standby for regeneration.

- Flow rate duty vessel = treated flow / 2

$$\frac{633 \text{ gpm}}{2} = 317 \text{ gpm}$$

Design Steps:

- Determine available equipment diameter using the linear velocity criteria:
  - Assume linear velocity: 11.2 gpm/ft^2
  - Vessel area = treated flow / linear velocity

$$\frac{317 \text{ gpm}}{11.2 \text{ gpm/ft}^2} = 28.3 \text{ ft}^2$$

  - Vessel diameter: $A = \pi r^2 \rightarrow d = 2\sqrt{\frac{A}{\pi}}$

$$28.3 \text{ ft}^2 = \pi r^2 \quad d = 2\sqrt{\frac{28.3 \text{ ft}^2}{\pi}} = 6 \text{ ft}$$

- Determine media volume per vessel based on specific volume criteria:
  - Assume specific flow rate: 2.9 gpm/ft^3

- Theoretical minimum vessel volume = treated flow / specific flow rate

$$\frac{317 \text{ gpm}}{2.9 \text{ gpm/ft}^3} = 109 \text{ ft}^3$$

- Determine volume of resin to achieve the minimum bed depth by varying the number of supersacks:
  - Assume resin is shipped in 42 ft^3 superacks (*ss*)

$$\frac{109 \text{ ft}^3}{42 \text{ ft}^3/\text{ss}} = 2.6\frac{\text{supersacks}}{\text{vessel}} \rightarrow 3\frac{\text{supersacks}}{\text{vessel}}$$

  - Vessel resin volume = 42 ft^3 × 3 = 126 ft^3 of resin per vessel.
- Determine depth of resin in vessel:
  - Vessel volume / vessel area = depth

$$\frac{126 \text{ ft}^3}{28.3 \text{ ft}^2} = 4.4 \text{ ft} \approx 4.5 \text{ ft}$$

- Determine EBCT
  - EBCT = vessel volume / treated flow rate

$$126 \text{ ft}^3 \times \frac{7.48 \text{ gal}}{\text{ft}^3} \times \frac{\text{min}}{317 \text{ gal}} = 2.9 \text{ min} \approx 3 \text{ min}$$

  - 3 min is within the EBCT goal criteria.

Design regeneration protocol.

- Determine backwash protocol:
  - Flow = vessel area × backwash linear velocity
    - Design basis for linear velocity is 2.4 gpm/ft^2.
    - $28 \text{ ft}^2 \times \frac{2.4 \text{ gpm}}{\text{ft}^2} = 68 \text{ gpm}$
  - Backwash duration = vessel volume / backwash flow rate

$$126 \text{ ft}^3 \times \frac{7.48 \text{ gal}}{\text{ft}^3} \times \frac{\text{min}}{68 \text{ gal}} = 14 \text{ min}$$

    - Assume 2 BV based on 2.4 gpm/ft^2 at 20°C

$$14 \text{ min} \times 2 = 28 \text{ min}$$

  - Total waste per step = flow × duration

$$28 \text{ min} \times 68 \text{ gpm} = 1{,}904 \text{ gal}$$

- Determine brine step
  - Flow = vessel volume × backwash specific flow rate
    - Design basis for specific flow rate is 0.5 gpm/ft³.
    - $126\ ft^3 \times \frac{0.5\ gpm}{ft^3} = 63\ gpm$
  - Brine step duration use design criteria of 30 min
  - Total waste per step = flow × duration

$$30\ min \times 63\ gpm = 1{,}890\ gal$$

- Determine slow rinse step:
  - Flow = vessel volume × backwash specific flow rate
    - Design basis for specific flow rate is 0.5 gpm/ft³.
    - $126\ ft^3 \times \frac{0.5\ gpm}{ft^3} = 63\ gpm$
  - Slow rinse duration use design criteria of 40 min
  - Total waste per step = flow × duration

$$40\ min \times 63\ gpm = 2{,}520\ gal$$

- Determine fast rinse step:
  - Flow = flow per vessel = 317 gpm
  - Fast rinse duration use design criteria of 20 min
  - Total waste per step = flow × duration

$$20\ min \times 317\ gpm = 6{,}340\ gal$$

Figure 9-3 shows a flow schematic for SBA co-flow regenerable configuration with bypass flow. In this case, Vessel 3 is off-line while in regeneration.

Table 9-6 presents a summary of the design criteria for the SBA regenerable system and Table 9-7 presents a summary of the waste volumes.

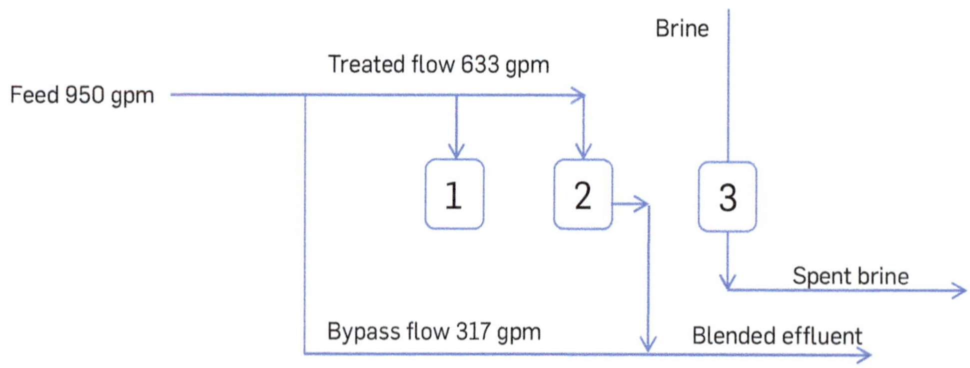

Figure 9-3 SBA regeneration co-flow regeneration Example: Vessel 3 is offline while in regeneration

Courtesy of Purolite Corp.

Table 9-6 SBA example design parameters

Design Parameters	Value
Total flow rate	950 gpm
Influent chrome VI	20 µg/L
Leakage chrome VI	2 µg/L
Effluent concentration goal (blended)	8 µg/L
Bypass flow percent	33.3%
Bypassed flow rate	317 gpm
Treated flow rate	633 gpm
Number of vessels in service	2
Number of vessels in standby	1
Number of vessels total	3
Flow per vessel	317 gpm
Vessel diameter	6 feet
Volume per supersack	42 ft^3
Number of supersacks per vessel	3
Media volume per vessel	126 ft^3
Media volume total onsite	378 ft^3
Vessel area	28 ft^2
Bed depth per vessel	4.59 feet
Linear velocity: 8–12 gpm/ft^2 target per vessel	11.2 gpm/ft^2
Specific flow rate: 1–5 gpm/ft^3 target per vessel	2.5 gpm/ft^3
EBCT	3.0 minutes
BV/hour	20.2
Estimated throughput, BV	9,925 BVs
Bed life	20.5 days

Table 9-7 Regeneration: Based on 10 lb/ft^3 NaCl and 10 percent brine solution

	Design Basis (gpm/ft^2)	Flow (gpm)	Time (min)	Total Waste per Step (gallons)
Backwash*	2.4 gpm/ft^2	68	28	1,904
Brine step	0.5 gpm/ft^3	63	30	1,890
Slow rinse	0.5 gpm/ft^3	63	40	2,520
Fast rinse		317	20	6,340
Totals			118	12,654

* Backwash based on using a typical Type I SBA resin. Influent water temperature 20°C, 50 percent backwash expansion occurs at approximately 2.4 gpm/ft^2. Time is based on backwashing 2 bed volumes.

REFERENCES

"Chromium | History, Uses, Facts, Physical & Chemical Characteristics." 2018. *Periodic Table* (blog). October 19, 2018. https://periodic-table.com/chromium/.

Edzwald, J.K. 2011. *Water Quality and Treatment: A Handbook on Drinking Water, Sixth Edition*. 6th ed. AWWA, American Society of Civil Engineers, McGraw-Hill. https://www.awwa.org/Store/{ProductName}/ProductDetail/{ProductId}?productId=6292.

Health Canada. 2018. "Guidelines for Canadian Drinking Water Quality: Guideline Technical Document - Chromium." Guidance. Aem. August 17, 2018. https://www.canada.ca/en/health-canada/services/publications/healthy-living/guidelines-canadian-drinking-water-quality-guideline-technical-document-chromium-profile.html.

Health Canada. 2015. "Chromium in Drinking Water." Consultations. Aem. July 22, 2015. https://www.canada.ca/en/health-canada/programs/chromium-drinking-water/chromium-drinking-water.html.

Jacobs. 2019. "Hexavalent Chromium Treatment with Strong Base Anion Exchange." Water Research Foundation. https://www.waterrf.org/research/projects/hexavalent-chromium-treatment-strong-base-anion-exchange.

USEPA (US Environmental Protection Agency). 2015. "Chromium in Drinking Water." Overviews and Factsheets. USEPA. September 22, 2015. https://www.epa.gov/sdwa/chromium-drinking-water.

USEPA. 2007. "Drinking Water Treatability Database." Drinking Water Treatability Database. March 2007. https://tdb.epa.gov/tdb/contaminant?id=1175876466.

WHO (World Health Organization). 1993. "Background Document for the Development of WHO Guidelines for Drinking-Water Quality.–2nded." *Volume* 2: 281–83.

CHAPTER 10

Per- and Polyfluoroalkyl Substances (PFAS) Removal

10.1 Introduction

Per- and polyfluoroalkyl substances (PFAS) are a family of several thousand man-made chemicals, the earliest of which were first introduced in the 1940s. They have been used in manufacturing and in industrial and consumer products. Due to their widespread use and resistance to breaking down, they have become ubiquitous in the environment. Detectable levels of the most common PFAS chemicals may be found not only in many source waters but also in the blood serum of most residents of industrialized nations, including the United States.

PFAS have become a contaminant of concern because studies indicate that exposure to certain PFAS can have negative health impacts. While additional research to better understand PFAS impacts on human health is ongoing, studies of perfluorooctanoic acid (PFOA) and perfluorooctanesulfonic acid (PFOS), the most widespread and researched PFAS compounds, suggest potential health effects ranging from reproductive, developmental, and immune system impacts to liver and kidney effects, tumor formation, and cancer (USEPA 2016).

This chapter describes how ion exchange and non-regenerable media systems can be used for the removal of PFAS in groundwater. This chapter describes:

- properties and chemistry associated PFAS,
- regulatory issues associated with PFAS,
- source water quality parameters that can impact the selection of ion exchange resins suitable for PFAS removal,
- an overview of bench and pilot testing requirements,
- design guidelines including calculations and residuals generation, and
- design example.

10.2 PFAS Occurrence and Chemistry

10.2.1 Occurrence and Uses

PFAS chemicals have been in use for decades in manufacturing and are found in many common products. They may be found in firefighting foams to extinguish fuel-based fires (aqueous film forming foams) and in everyday products including non-stick cookware, stain resistant and water repellant fabrics, personal care products, food packaging, paints, and waxes. Due to the physical and chemical properties of PFAS chemicals, they do not break down readily and persist in the environment. The use of firefighting foams at airports, military bases, and other firefighting training facilities have become a significant source of PFAS contamination in water supplies (USEPA 2016).

Current treatment processes at drinking water, wastewater and landfill leachate facilities are often not effective at removing PFAS. PFAS have been able to travel long distances from their original sources. They can be found in many bodies of water and ground water supplies, exposing humans and animals to their bioaccumulative effects. While the use of PFOA and PFOS, the two most widely used PFAS chemicals, have been phased out in the United States because of concerns about their health effects, certain replacement chemicals such as GenX (replacement for PFOA), which were originally believed to be safer, are now also contaminants of concern.

Drinking water sources where PFAS contamination is present typically have contamination at ng/L (ppt) concentration levels. Commercial laboratories continue to expand the number of PFAS chemicals that can be analyzed at increasingly lower detection limits. With current testing capabilities, PFOA and PFOS are the most common PFAS, but it is not uncommon to detect additional PFAS compounds, depending on the manufacturing or contamination sources in the region of the water supply. For this reason, in addition to PFOA and PFOS, several other PFAS compounds are the subject of drinking water health-based guidelines and rulemaking in individual states, including:

- Perfluorononanoic acid (PFNA),
- Perfluorohexane sulfonic acid (PFHxS),
- Perfluoroheptanoic acid (PFHpA),
- Perfluorobutane sulfonic acid (PFBS),
- Perfluorobutanoic acid (PFBA),
- Perfluorodecanoic acid (PFDA), and
- Hexafluoropropylene oxide dimer acid (HFPO-DA or GenX).

10.2.2 Chemistry

A defining characteristic of PFAS chemicals is the presence of carbon-fluorine bonds, which is the strongest bond in organic chemistry. The properties of these bonds make PFAS chemicals difficult to break down fully under normal environmental conditions and have earned them the nickname "forever chemicals." There

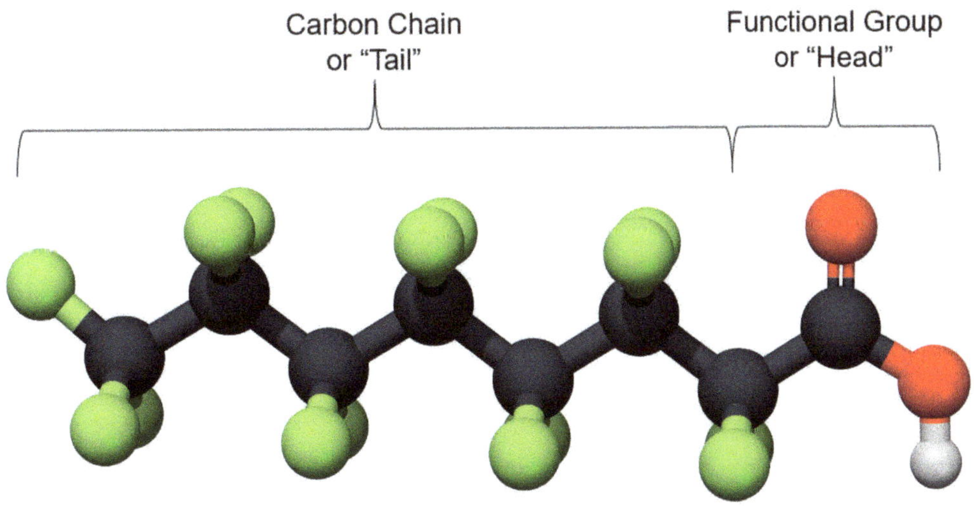

Figure 10-1 Diagram of PFOA molecule

are two major groups of PFAS chemicals: *per*fluoroalkyl compounds and *poly*fluoroalkyl compounds. Perfluoroalkyl compounds have carbon-fluorine bonds as their backbone and are considered terminal PFAS compounds that do not break down further. Polyfluoroalkyl compounds have perfluoroalkyl components in addition to non-fluorinated components like carbon-hydrogen bonds that can break down under certain environmental conditions. When polyfluoroalkyl compounds break down, they break down into perfluoroalkyl compounds that do not break down any further.

Figure 10-1 presents the typical structure of a simple linear perfluoroalkyl acid (PFAA), a common subgroup of perfluoroalkyl compounds. The chemical represented in the figure is the well-known PFAA compound PFOA, with the gray spheres representing carbon atoms, the green spheres representing fluorine atoms, the red spheres representing oxygen atoms and the white sphere representing hydrogen. The molecule has two components, the carbon chain or "tail" with the carbon-fluorine bond backbone and the functional group or "head." The tail portion of the molecule is hydrophobic, whereas the head is hydrophilic and takes the form of an anion when the hydrogen dissociates. Due to these hydrophobic and hydrophilic properties, like a surfactant, many of these compounds tend to prefer an air water interface.

Common PFAS compounds are often referred to as long chain or short chain and this is in reference to the number of carbon atoms in the perfluoroalkyl carbon chain. Longer chain molecules have a longer half-life in human bodies. It can take on the order of years for longer chain PFAS to be eliminated from the body since last exposure. Shorter chain PFAS are typically eliminated from the body in a timeframe of months. Longer chain PFAS, like PFOA and PFOS, with concerns about greater bioaccumulation and toxic effects, have historically been the focus of health-based research and reduction efforts. Refer to Table 10-1 for a list of common short and long chain PFAS.

Table 10-1 Common PFAS compounds

Acronym	Name	Molecular Formula
Short Chain PFCAs		
PFBA	Perfluorobutanoic acid	C_3F_7COOH
PFPeA	Perfluoropentanoic acid	C_4F_9COOH
PFHxA	Perfluorohexanoic acid	$C_5F_{11}COOH$
PFHpA	Perfluoroheptanoic acid	$C_6F_{13}COOH$
Long Chain PFCAs		
PFOA	Perfluorooctanoic acid	$C_7F_{15}COOH$
PFNA	Perfluorononanoic acid	$C_8F_{17}COOH$
PFDA	Perfluorodecanoic acid	$C_9F_{19}COOH$
PFUnA or PFUnDA	Perfluoroundecanoic acid	$C_{10}F_{21}COOH$
PFDoA or PFDoDA	Perfluorododecanoic acid	$C_{11}F_{23}COOH$
PFTrDA	Perfluorotridecanoic acid	$C_{12}F_{25}COOH$
PFTeDA or PFTA	Perfluorotetradecanoic acid	$C_{13}F_{27}COOH$
Short Chain PFSAs		
PFBS	Perfluorobutane sulfonic acid	$C_4F_9SO_3H$
PFPeS	Perfluoropentane sulfonic acid	$C_5F_{11}SO_3H$
Long Chain PFSAs		
PFHxS	Perfluorohexane sulfonic acid	$C_6F_{13}SO_3H$
PFHpS	Perfluoroheptane sulfonic acid	$C_7F_{15}SO_3H$
PFOS	Perfluorooctane sulfonic acid	$C_8F_{17}SO_3H$
PFNS	Perfluorononane sulfonic acid	$C_9F_{19}SO_3H$
PFDS	Perfluorodecane sulfonic acid	$C_{10}F_{21}SO_3H$
PFUnS or PFUnDS	Perfluoroundecane sulfonic acid	$C_{11}F_{23}SO_3H$
PFDoS or PFDoDS	Perfluorododecane sulfonic acid	$C_{12}F_{25}SO_3H$
PFTrDS	Perfluorotridecane sulfonic acid	$C_{13}F_{27}SO_3H$
PFTeDS	Perfluorotetradecane sulfonic acid	$C_{14}F_{29}SO_3H$
Fluorotelomer Sulfonic Acids		
4:2 FTS	1H,1H,2H,2H-Perfluorohexane sulfonic acid	$C_4F_9CH_2CH_2SO_3H$
6:2 FTS	1H,1H,2H,2H-Perfluorooctane sulfonic acid	$C_6F_{13}CH_2CH_2SO_3H$
8:2 FTS	1H,1H,2H,2H-Perfluorodecane sulfonic acid	$C_8F_{17}CH_2CH_2SO_3H$

(continued)

Table 10-1 Common PFAS compounds (*continued*)

Acronym	Name	Molecular Formula
Ether Carboxylic and Sulfonic Acids		
HFPO-DA (GenX)	Hexafluoropropylene oxide dimer acid	Not available
ADONA	4,8-dioxa-3H-perfluorononanoic acid	Not available
9Cl-PF3ONS	9-chlorohexadecafluoro-3-oxa-none-1-sulfonic acid	Not available
11Cl-PF3OUdS	11-Chloroeicosafluoro-3-oxaunde-cane-1-sulfonic acid	Not available
PFMBA	Perfluoro-4-methoxybutanoic acid	Not available
PFMPA	Perfluoro-3-methoxypropanoic acid	Not available
NFDHA	Nonafluoro-3,6-dioxaheptanoic acid	Not available
PFEESA	Perfluoro (2-ethoxyethane) sulfonic acid	Not available
Perfluorooctane sulfonamidoacetic acids		
NEtFOSAA	N-ethyl perfluorooctanesulfonamidoacetic acid	$C_8F_{17}SO_2N(C_2H_5)CH_2COOH$
NMeFOSAA	N-methyl perfluorooctanesulfonamidoacetic acid	$C_8F_{17}SO_2N(CH_3)CH_2COOH$

The group of PFAS chemicals that are most commonly detected in the environment are PFAAs and some of their more widely used replacements like GenX and common intermediate products. PFAAs include two major subgroups: perfluoroalkyl carboxylic acids (PFCAs) and perfluoroalkyl sulfonic acids (PFSAs). PFCAs have carbon chains of varying lengths and the -COOH functional group. PFSAs likewise have carbon chains of varying lengths and the functional group $-SO_3H$. PFAAs are thermally and chemically stable and do not degrade under normal environmental conditions. PFAA contaminants in water sources are found in their dissociated anionic state (ITRCA 2020).

10.3 PFAS Regulations

10.3.1 United States Regulations and Guidelines

In May 2016, the US Environmental Protection Agency (USEPA) announced the release of lifetime health advisories and health effects supporting documents for PFOA and PFOS. The USEPA's health advisories, which identify the concentration of PFOA and PFOS in drinking water which adverse health effects are not anticipated to occur over a lifetime of exposure, are 70 ng/L for PFOA and PFOS combined. Health advisories are not enforceable and reflect the USEPA's assessment of the best available peer-reviewed science. These health advisories superseded the USEPA's 2009 provisional health advisories for PFOA and PFOS.

In June 2018, the Agency for Toxic Substances and Disease Registry (ATSDR) issued a draft Toxicological Profile for Perfluoroalkyls for public comment that indicated that the USEPA health advisories for PFOA and PFOS should potentially be lowered.

On January 19, 2021, the EPA announced the final regulatory determinations to regulate PFOA and PFOS, and the process for establishing a national primary drinking water regulation for PFOA and PFOS is underway at the time of this writing. The proposed fifth Unregulated Contaminant Monitoring Rule (UCMR 5) was also announced in January 2021 and would require water utilities to collect samples for 29 PFAS compounds between 2023 and 2025. The PFAS occurrence data collected through the UCMR 5 process may lead to future determinations to regulate additional PFAS compounds.

Several states have established more stringent health-based guidance and advisories for PFAS compounds than the USEPA health advisories, as they consider formal regulation of PFAS contaminants at the state level for PFOA, PFOS, and other PFAS compounds. Several states are planning to adopt or have adopted MCLs for common PFAS compounds in advance of federal regulatory action. These MCLs are enforceable and require the implementation of a treatment system. States are generally targeting PFOA and PFOS levels that are more stringent than the USEPA health advisories and cover additional PFAS chemicals.

10.3.2 International Regulations and Guidelines

As of this writing, the World Health Organization does not have drinking water guidelines related to PFAS compounds. Several individual countries have health based values including Australia, Canada, Denmark, Germany, Italy, Netherlands, Sweden, and the United Kingdom. Depending on the country, these values may or may not be enforceable.

10.4 Ion Exchange Resin Treatment Options

There are several technologies that are currently commercially available for PFAS removal including activated carbon (granular and powdered), ion exchange and reverse osmosis. The technology selected for a particular PFAS removal application will depend on a number of considerations including PFAS contaminant levels, background water chemistry, potential other co-contaminants and sources of fouling, treatment goals, options for integration into an existing treatment train, site constraints, and waste handling considerations. This section focuses on treatment options using ion exchange resin technology.

Ion exchange resins that are suitable for PFAS removal are PFAS-selective strong base anion exchange resins. These resins can remove over 99.99 percent of PFAS chemicals that are commonly found in water supplies. In drinking water applications, these resins are currently available as single-use resins (often referred to as "PFAS selective resins").

PFAS selective resins can achieve high removal rates because they are designed to be more selective for the most common PFAS contaminants than other major anions typically present. The following is a typical order of selectivity for PFAS selective resins for several PFAS compounds in relation to other PFAS compounds as well as other common ions. PFAS compounds are shown in bold.

11Cl-PF3OUdS > **NEtFOSAA** > **9Cl-PF3ONS** > **NMeFOSAA** > **PFOS** > **PFHxS** > **PFBS** > **PFTrDA** > **PFTA** > **PFDoA** > **PFUnA** > **PFDA** > **PFNA** > **PFOA** > **ADONA** > **HFPO-DA** > **PFHpA** > ClO_4^- > **PFHxA** > **PFPeA** > **PFBA** > NO_3^- > SO_4^{2-} > Cl^- > HCO_3^- > F^-

These types of resins are generally more selective for the sulfonate than the carboxylate PFAS and have greater removal capacity for long chain PFAS than shorter chain PFAS compounds. These resins can remove PFAS not only by ion exchange but also by adsorption. The negatively charged head of PFAS compounds allow removal by ion exchange, but the tail of the compound also allows removal by adsorption particularly for longer chain PFAS. These dual removal mechanisms, along with rapid kinetics and high selectivity, allow PFAS selective resins to have high PFAS removal capacities using smaller volumes of media and contact times than other PFAS treatment options like granular activated carbon (GAC).

The most common source water quality parameters that impact the performance of ion exchange resins are the anions nitrate, sulfate, chloride, and bicarbonate alkalinity as well as total organic carbon. These competitive anions play a large role in resin capacity because they are in the water at roughly one million times higher concentrations than PFAS. Even though the resin is selective for PFAS, the competitive anions are the main driver for throughput. Table 10-2 summarizes important background water quality of interest for PFAS selective ion exchange resin treatment design. Providing resin manufacturers and suppliers this background water quality data along with PFAS contaminant levels can help the ion exchange system designer determine the type of resin, the system sizing criteria, and pretreatment requirements for a given application.

10.5 Bench and Pilot Testing

Chapter 13 of this handbook discusses the general process of piloting an ion exchange system. This section focuses on the specific considerations for PFAS

Table 10-2 PFAS ion exchange design background water quality

Parameter	Units
Sulfate	mg/L
Nitrate	mg/L as N or NO_3
Alkalinity	mg/L as $CaCO_3$
Chloride	mg/L
Fluoride	mg/L
Perchlorate	µg/L
Arsenic	µg/L
Chromium	µg/L
Uranium	µg/L
Calcium	mg/L as Ca
Magnesium	mg/L as $CaCO_3$
Sodium	mg/L
Potassium	mg/L
Iron (total and dissolved)	mg/L
Manganese (total and dissolved)	mg/L
pH	SU
ORP	mV
TDS	mg/L
Suspended solids	mg/L
Oil & grease	mg/L
Total organic carbon	mg/L
VOCs	µg/L

removal. While some resin manufacturers and suppliers have the ability to perform computer modeling to estimate resin full-scale performance using select water quality data, it may be beneficial to perform pilot testing for a number of reasons. Pilot testing provides an opportunity to screen and compare different types of PFAS selective resins for a given application and can help confirm or establish full-scale design parameters including surface loading rate, throughput, empty bed contact time (EBCT), waste characterization, and pretreatment requirements.

As with all bench and pilot testing, the test should mimic and be scalable to the anticipated full-scale system. The pilot testing apparatus generally has multiple columns that allow the comparison of different resin types and EBCTs. Due to the high affinity of the resin for PFAS, breakthrough of PFAS contaminants is on the order of several months at shorter EBCTs and the duration of the pilot test should be coordinated with the regulators and resin suppliers.

A well-designed pilot study will start with an extensive water quality analysis that includes testing for the suite of PFAS chemicals that may be analyzed at commercial laboratories as well as background water quality outlined in Table 10-2. This water quality data can help resin suppliers validate estimates of resin throughput for target PFAS contaminants.

Modified rapid small-scale column tests (RSSCT), a bench-scale testing method to obtain media breakthrough data in an accelerated timeframe, have been performed with PFAS selective resins, although there is no industry standard for this testing with resin. More development is needed to validate the RSSCT versus full scale performance.

10.6 Design Guidelines

Recommended design guidelines for PFAS selective ion exchange resins are described in this section. PFAS selective ion exchange resins for drinking water applications are single-use. Well-designed treatment systems will maximize PFAS removal capacity of the resin before requiring changeout. A well-designed treatment system not only implements best practices for resin system sizing and equipment configuration, but also provides pretreatment as needed and considers operation and maintenance aspects in the overall design, including ancillary equipment like feed pumps.

Since PFAS resins are single-use, they are not typically backwashed and therefore pretreatment is especially important. The primary pretreatment that is particularly important for single-use resins is suspended solids removal. Solids buildup on the resin bed can lead to uneven flow distribution and loading within the bed due to blockage and channeling. Typically, suspended solids removal is achieved by providing prefiltration using bag filters or cartridge filters.

Iron and manganese are common co-contaminants in PFAS removal applications, particularly in groundwater sources. While low levels of iron and manganese that remain in solution can pass through the resin system without a significant impact, pretreatment should be considered, particularly if it is determined that particulate iron or manganese are present (Purolite n.d.).

Refer to Chapter 2: Fundamentals of Ion Exchange, Section 2.3, for more detailed description of water quality pretreatment goals and options.

The following are general design guidelines for the PFAS selective ion exchange resin system sizing, and these should be customized based on the specific project requirements:

- EBCT: 1.5–3 min per vessel
- Hydraulic loading rate: 6–18 gpm/ft^2 (15–45 m/h)
- Resin depth: 3 ft minimum at loading rates up to 12 gpm/ft^2; 3.7 ft at loading rates greater than 12 gpm/ft^2

- Specific flow rate: 1–5 gpm/ft^3 (8–40 bed volume [BV]/h)
- Configuration: lead/lag vessel (series) configuration
- Minimum pretreatment: solids removal to 5 or 10 microns

Some PFAS selective ion exchange resin applications involve the conversion of existing GAC vessel infrastructure to ion exchange systems. The design guidelines and best practices outlined above apply to such systems. Since GAC vessels are typically operated in a lower range of liquid loading rates, the conversion of GAC vessels to ion exchange vessels may lead to lower than recommended hydraulic loading rates for resin for the range of design flows. Options to address potential lower loading rate impacts on resin performance should be reviewed with the resin supplier and equipment supplier.

Please see Chapter 2: Fundamentals of Ion Exchange for general equipment and guidance.

10.7 Disposal Options

The PFAS selective single-use ion exchange resin system generates two potential wastewaters: backwash waste and rinsing waste primarily associated with resin system startup and resin changeouts. Once the new resin is placed into continuous service, no wastewater is generated as part of normal operation.

Backwashing: The purpose of initial resin backwashing is typically for the removal of resin fines that may be present in the delivered resin and grading of the resin bed as much as possible. The recommendations for initial backwashing will vary depending on the resin supplier and the resin's physical properties. Some suppliers recommend backwashing for 30–90 min at as much as 50 percent resin bed expansion, while others take the position that backwashing is not necessary. It is good practice to include provisions in the design for backwashing the resin. Even if the resin that is initially installed does not require backwashing, this may change if in the future a different resin is used.

Recommended backwash rates will vary depending on the water temperature and a given resin's properties, but typical backwash rates to achieve 50 percent bed expansion at 55°F are approximately 1–2 gpm/ft^2. The water source for resin backwashing should be considered in the design. If chlorinated water sources (e.g., finished water) are proposed, the impact of the short-term oxidant exposure on the resin should be reviewed with the resin supplier. If direct sewer discharge at the desired backwash rates is possible, backwash waste handling may be straightforward. If direct discharge is not possible or if sewer is not available, provisions for temporary or permanent waste handling infrastructure should be provided.

In general, once placed into service, backwashing of the bed is not recommended, except as a last resort to try to alleviate a resin solids build-up or clogging issue that would otherwise require premature changeout of a resin bed. Once resin has been placed into service, it will begin to develop zones in the bed with the top of the bed characterized by spent resin, the bottom of the bed characterized by "fresh" resin, and the area in between acting as a mass transfer zone for the target contaminant, which slowly descends through the bed as the resin capacity is exhausted. Backwashing of the bed once in service will disturb the mass transfer zone and potentially lead to premature contaminant breakthrough.

Rinsing: PFAS selective ion exchange resins typically need an initial forward flow rinse prior to use in potable water applications. Most resins have an initial minimum rinsing requirement to meet NSF 61 certification requirements that can sometimes be performed prior to delivery. Where rinsing will not be performed prior to delivery, a plan for temporary or permanent infrastructure for waste handling will need to be considered where direct discharge to a sewer is not possible. A resin application where the resin is not pre-rinsed prior to delivery will typically require a minimum rinse of 20 BV. The required minimum BV for initial rinsing should be provided by the resin supplier.

PFAS selective ion exchange resins delivered in an unbuffered form will temporarily impact water quality before the resin reaches equilibrium with the background water chemistry. Figure 10-2 is an example of the short-term impacts on background water chemistry over the first 480 BV of a fresh chloride form PFAS selective resin bed. Resin behavior in an application will depend on the type of resin and the background water chemistry.

For this example, initially there are significant changes in major background anions as the chloride anions on the fresh resin are exchanged for anions that the resin prefers in the feedwater. In this example, the most significant water quality impact occurs during the first 100 BV, and resin equilibrium is achieved at approximately 440 BV. A customer may want or need to perform additional rinsing to overcome these initial water quality impacts to achieve the desired effluent water quality and alleviate potential concerns regarding negative distribution system water quality impacts. Provisions for handling this additional rinsing waste should be provided if needed. Use of pre-buffered forms of PFAS selective resins, while more expensive, may significantly reduce these background water quality impacts.

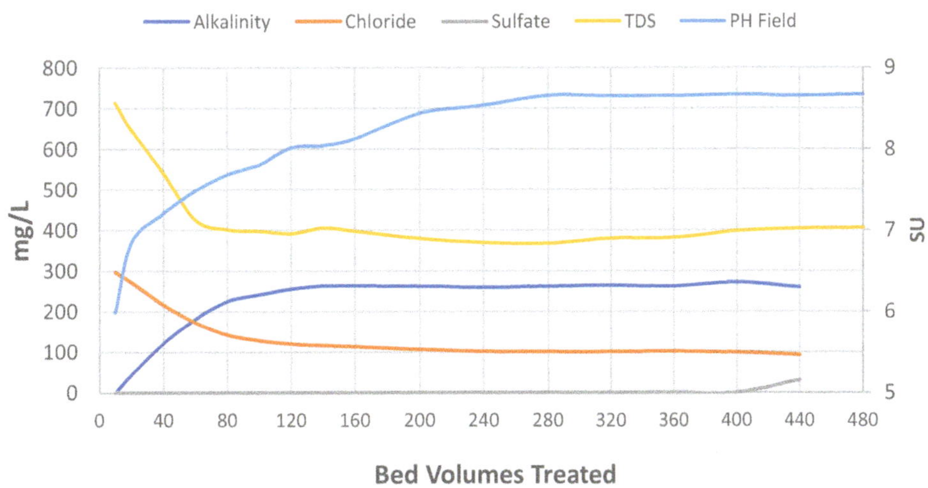

Figure 10-2 PFAS selective resin standard chloride form initial rinse water quality

10.7.1 Resin Disposal Options

PFAS selective single-use resins will eventually become exhausted and require replacement. Planning for spent resin disposal should begin at the design development stage of the project and should be re-evaluated once the ion exchange system is commissioned and again during system operation. There are currently two potential disposal options, landfill disposal and incineration at high temperatures. Factors to consider with respect to resin disposal include:

- characterization of the spent resin,
- local, regional, and national regulations regarding landfill disposal,
- local, regional, and national regulations regarding incineration,
- equipment features to remove the spent resin,
- safety requirements with respect to handling of the spent resin, and
- current and future liability of the utility and potential methods to reduce this liability.

Regulations regarding PFAS continue to change. Due to these changes a periodic review of these issues will assist the utility in determining the optimum disposal options. Coordination with the resin suppliers, waste disposal companies and regulators should be performed when evaluating waste disposal options.

10.8 Design Example

This design example is for treatment of a groundwater well source for drinking water. PFAS have been detected in the well source at levels that exceed the USEPA's health advisory for PFOA and PFOS of 70 ng/L combined. Table 10-3 summarizes raw water quality parameters.

The PFAS treatment goal for this water source is to treat to below the detection limit for PFOA and PFOS. Given that PFAS selective resins are more selective for sulfonates (e.g., PFOS) than carboxylates (e.g., PFOA), PFOA is expected to break through the resin before PFOS. A throughput projection can be done by contacting the resin supplier or through pilot testing. In this example, a pilot test was conducted that indicated a minimum of 150,000 BV treated would be achieved prior to PFOA breakthrough. The raw water quality does have specific characteristics that cause concern with respect to fouling. A 5-micron cartridge filter pretreatment system will be provided for solids removal to protect the resin bed from solids buildup over the life of the resin. Table 10-4 presents the resin system design parameters for this PFAS treatment design example, and Figure 10-3 presents a process flow diagram.

Design Steps:

- Determine ion exchange resin vessel diameter:
  - Assume single train consisting of two (lead/lag configured) 8 ft diameter vessels: 2 × radius, r = 4 ft
  - Determine vessel cross sectional area: $A = \pi r^2 = 50.3\ \text{ft}^2$

Table 10-3 Water source parameters

Raw Water Parameter	Value
Design flow rate	550 gpm
Average flow rate	500 gpm
Minimum flow rate	400 gpm
PFOA	30 ng/L
PFOS	55 ng/L
PFNA	< 2 ng/L (non-detect)
PFHxS	25 ng/L
PFHpA	< 2 ng/L (non-detect)
PFBS	6 ng/L
Alkalinity	150 mg/L as $CaCO_3$
Chloride	120 mg/L
Sulfate	25 mg/L
Nitrate	2.5 ppm as N
Uranium	<1 µg/L
pH (field)	7.2
Calcium	70 mg/L
Iron, total	0.03 mg/L
Manganese, total	0.01 mg/L
TOC	1.0 mg/L
TDS	300 mg/L

Table 10-4 PFAS selective ion exchange system design parameters

Design Parameters	Value
Design flow rate	550 gpm
Average flow rate	500 gpm
Minimum flow rate	400 gpm
Influent PFOA	30 ng/L
Influent PFOS	55 ng/L
Effluent concentration goal	<2 ng/L PFOA and PFOS (each)
Number of trains	1
Number of vessels per train	2
Vessel diameter	8 ft
Design flow per vessel	550 gpm
Minimum flow per vessel	400 gpm
Volume per supersack	42 ft^3
Number of supersacks per vessel	5
Media Volume per vessel	210 ft^3
Media Volume total onsite	420 ft^3

(continued)

Table 10-4 PFAS selective ion exchange system design parameters (*continued*)

Design Parameters	Value
Vessel area	50.3 ft^2
Bed depth	4.2 ft
Linear velocity rate: goal 8–12 gpm/ft^2:	10.9 gpm/ft^2
Specific flow rate: 1–5 gpm/ft^3 target per vessel	2.6 gpm/ft^3
EBCT goal: 2–3 min	2.9 minutes
Projected throughput	150,000 BV
Estimated days between lead vessel exchanges	327 days at average flow rate

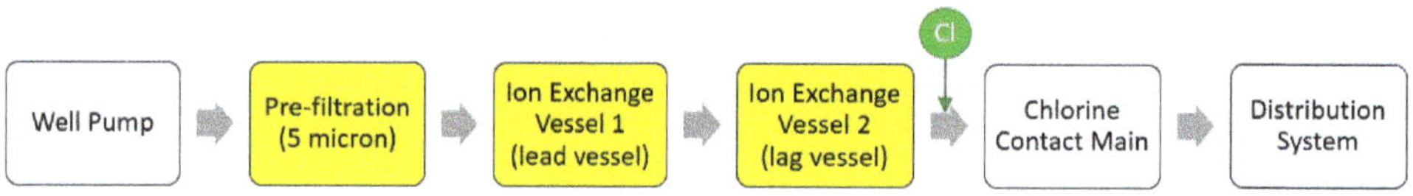

Figure 10-3 PFAS treatment design example process flow diagram

- Calculate design linear velocity:
  - Design flow rate / vessel cross sectional area = 550 gpm/ 50.3 ft^2 = 10.9 gpm/ft^2
  - 10.9 gpm/ft^2 falls into the recommended range of 6–18 gpm/ ft^2 (15–45 m/h)
- Calculate minimum linear velocity:
  - Minimum flow rate / vessel cross sectional area = 400 gpm/ 50.3 ft^2 = 8.0 gpm/ft^2
  - 8.0 gpm/ft^2 falls into the recommended range of 6–18 gpm/ft^2 (15–45 m/h)

- Determine volume of resin to achieve the desired minimum EBCT by varying the number of supersacks. It is much easier to load and unload based on using full supersack volumes.
  - Target minimum EBCT per vessel: 2.5 min
  - Calculate minimum resin volume per vessel:
    - Minimum EBCT × design flow rate = 2.5 min × 550 gpm/ 7.48 gal/ft^3 = 184 ft^3 resin per vessel (minimum)
    - Minimum resin volume per vessel / supersack volume = 184 ft^3 resin / 42 ft^3 = 4.4 supersacks
    - Use 5 supersacks per vessel: 42 ft^3 per supersack × 5 supersacks = 210 ft^3 resin per vessel
  - Check bed depth:
    - Resin volume per vessel / vessel cross sectional area = 210 ft^3 / 50.3 ft^2 = 4.2 ft bed depth
    - This meets the minimum bed depth of 3 ft.

  - Calculate EBCT per vessel at design flow rate:
    - Resin volume per vessel × gallons per cubic foot / design flow rate = 210 ft^3 resin × 7.48 gal/ft^3 / 550 gpm = 2.9 min
- Calculate the days between exchanges based on estimated 150,000 BV until PFOA breakthrough:
  - Calculate one vessel BV in gallons: 210 ft^3/BV × 7.48 gal/ft^3 = 1,571 gal/BV
  - Calculate 150,000 BV in gallons: 1,571 gal/BV × 150,000 BV = 235,650,000 gal treated
  - Calculate number of days to treat 150,000 BV: 235,650,000 gal/ 500 gpm (average flow rate) / 1,440 min/day = 327 days

Residuals Handling Considerations:

- Determine backwash requirements if backwash is deemed necessary:
  - Calculate backwash rate assuming maximum backwash rate of 1.5 gpm/ft^2:
    - Vessel cross sectional area × backwash rate = 50.3 ft^2 × 1.5 gpm/ft^2 = 75 gpm
  - Calculate backwash volume per vessel assuming a maximum backwash duration of 90 min
    - Backwash rate × 90 min = 75 gpm × 90 min = 6,750 gal
- Determine rinsing requirements, if rinsing of installed resin is required:
  - Assume the maximum rinse rate will be the system design capacity of 550 gpm.
  - The minimum rinse rate will be 6 gpm/ft^2 to maintain adequate flow distribution through the bed, or a calculated flow rate of 50.3 ft^2 × 6 gpm/ft^2 = 302 gpm.
  - Calculate initial rinse volume assuming 20 BV initial rinse requirement:
    - Resin bed volume per vessel × initial rinse requirement = 1,571 gal/BV × 20 BV = 31,420 gal per vessel
  - Calculate potential extended rinse volume per vessel assuming 100 BV will achieve minimum effluent water quality desired for this application:
    - Resin bed volume per vessel × extended rinse requirement = 1,571 gal/BV × 100 BV = 157,100 gal per vessel
- Review residuals generation rates and volumes to determine options for handling and disposal and whether the cost of buffered resin should be considered for the application.

REFERENCES

ITRCA. 2020. "Technical/Regulatory Guidance: Per- and Polyfluoroalkyl Substances (PFAS)." The Interstate Technology & Regulatory Council.

Purolite. n.d. "PFAS-Selective Single Use Ion Exchange Resin for Drinking Water Systems." Application Guide. Design and Operational Guidelines. https://www.purolite.com/dam/jcr:e7dc9abb-abeb-4363-b46a-e44f1ae4ce18/PFAS-Selective%20Single%20Use%20Ion%20Exchange%20Resin%20for%20Drinking%20Water%20Systems_Application%20Guide.pdf.

USEPA (US Environmental Protection Agency). 2016. "Basic Information on PFAS." Overviews and Factsheets. USEPA. March 30, 2016. https://www.epa.gov/pfas/basic-information-pfas.

CHAPTER 11
Ion Exchange for Potable Water Reuse

11.1 Introduction

Water scarcity is a growing global concern with population growth, urbanization, pollution, drought and climate change exerting pressures on conventional water resources (WHO 2017). Water reuse is a strategy to mitigate the impacts of water scarcity by capturing wastewaters and treating them for use in a range of applications including potable water supply.

Potential water reuse feed sources include industrial effluent, produced water from oil and gas extraction, stormwater or mine water (Wong, Alspach, and Chalmer 2018). Final effluent from municipal sewage treatment works represents the most abundant source of feedwater for municipal reuse and is the main resource considered in this chapter.

Municipal wastewater for potable reuse requires advanced treatment. Membrane processes such as reverse osmosis (RO) and ultra-filtration (UF) are currently the dominant treatment techniques for water reuse. Other processes such as ozonation, activated carbon adsorption, ultraviolet irradiation, advanced oxidation, nitrogen polishing, and product water stabilization are also typically required to address specific contaminants depending on the characteristics and nature of the wastewater being treated (Wong, Alspach, and Chalmer 2018).

Ion exchange does not feature prominently in currently installed potable water reuse schemes. Indeed in 2017 the USEPA Potable Reuse Compendium stated that ion exchange is not currently used in any potable water reuse applications (USEPA 2017). Nevertheless, the development of increasingly stringent water reuse drinking water quality standards and a significant increase in the ability to detect microconstituent chemicals may open opportunities for ion exchange to address specific contaminants as part of a multi-barrier approach. The development of additional regulatory standards for water reuse is a long term process and often laboratory methods to detect microconstituent chemicals is proceeding at a faster pace than the regulatory standards. Utilities may wish to consider a "prevention" approach where, in the absence of regulatory standards, the removal of microconstituents may be beneficial to consider.

This chapter provides a brief introduction to potable water reuse, discusses general regulatory considerations related to water reuse and highlights some possible applications for ion exchange in water reuse systems.

11.2 Potable Water Reuse

Potable water reuse is typically divided into two main categories: direct potable reuse (DPR) and indirect potable reuse (IPR) (see Figure 11-1).

IPR occurs where reclaimed water is passed initially to an 'environmental buffer' such as an aquifer, surface water reservoir or river where it is blended with natural water before being abstracted for potable water treatment. IPR treatment plants are operational in the United States, Europe, Singapore, India, South Africa, Mexico and Australia (USEPA Potable Reuse Compendium 2017). The environmental buffer in IPR enables attenuation of flows and allows time to detect and respond to failures before distribution to supply. It may also assist with public acceptance of treated wastewater being returned for potable use (WHO 2017).

DPR differs from IPR in that reclaimed water is delivered directly to a drinking water treatment plant or municipal distribution network without first passing through an environmental buffer (see Figure 11-1).

IPR is much more widely used than DPR; however, a small number of DPR schemes have been completed. These include:

- Windhoek, Namibia, (5.5 mgd) which was upgraded in 1997 with reclaimed water being blended with conventionally treated surface water. This plant can provide around 50 percent of the city's potable water demand (USEPA Potable Reuse Compendium 2017).
- Big Spring in Texas, United States, (1.8 mgd) where reclaimed water is blended with raw surface water for potable reuse has been operational since 2013.
- Wichita Falls, Texas, (5.0 mgd) where a temporary DPR plant began operation in 2014 in response to extreme drought.

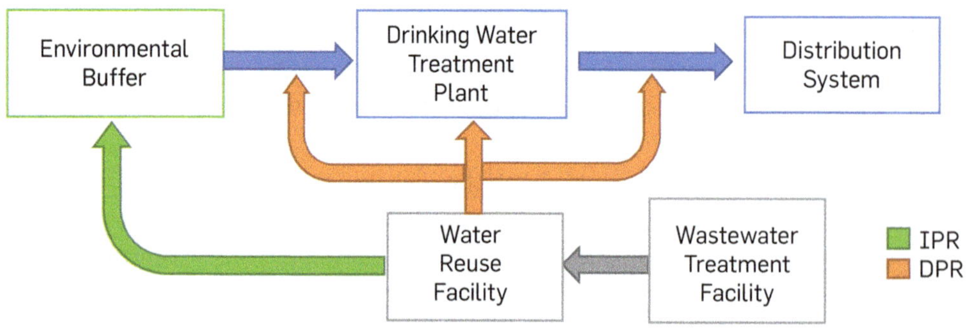

Figure 11-1 Possible blending locations representing DRP and IPR (based on WHO 2017)

11.3 Municipal Wastewater Hazards and Health Risks

Municipal wastewater contains a broad array of constituents which are not normally found in conventional potable water sources. These may be of industrial or domestic origin and may include regulated and unregulated pathogens and chemical contaminants.

Examples of potential contaminants in wastewater include disinfection by products, heavy metals, pharmaceuticals, pesticides, volatile organic carbons (VOCs), dioxins, polychlorinated biphenyls (PCBs), per- and polyfluoroalkyl substances (PFAS), personal care products, steroid hormones, endocrine disrupting compounds, engineered nano-particles, nutrients and radionuclides. Contaminants of emerging concern (CEC) also continue to be discovered.

Due to the range and concentration of potential contaminants, municipal wastewater offers a significant treatment challenge compared to conventional water sources. To compound difficulties, the composition of municipal wastewater may change (seasonally and diurnally), or with changes to activity in the service area such as the installation of a new industrial facility. Furthermore, new products are being introduced which will often be discharged to the wastewater system. Removing such an array of complex contaminants usually requires advanced treatment and a multiple barrier approach to ensure the safety and reliability of potable water supply.

11.4 Potable Water Reuse Regulations and Guidance

United States

Drinking water quality in the United States is governed by the Safe Drinking Water Act (SDWA) which was passed in 1974 and amended in 1986 and 1996. This gives authority to the US Environmental Protection Agency (USEPA) to set national drinking water quality standards.

Direct or indirect potable water reuse schemes must, as a minimum, meet the drinking water standards set in the SDWA requirements. Individual states can also set and enforce their own drinking water standards provided they are at least as stringent as the national standards (USEPA). For IPR and DPR, individual state requirements are often more stringent than the drinking water regulations for example in California, regulations adopted in 2014 for the replenishment of groundwater require pathogen reduction requirements of 12-log for enteric virus, 10-log for *Cryptosporidium* oocyst and 10-log for *Giardia* cyst (WHO 2017).

Constituents of emerging concern (CECs) such as endocrine disrupting compounds, 1,4 dioxane and NDMA, though not currently regulated for drinking water facilities must also be monitored with treatment steps provided for their removal (Wong, Alspach, and Chalmer 2018).

When reclaimed water is first passed to an environmental buffer (IPR), drinking water source and environmental protection regulations also apply. The principal regulation regulating discharges into surface waters is the Clean Water

Act (CWA). In addition, the SDWA also sets requirements for states to identify potential sources of drinking water contamination.

The 2012 Guidelines for Water Reuse contain recommendations for IPR and give information on applicable treatment processes, water quality criteria and monitoring requirements. These Guidelines also recommend that environmental buffers provide at least a two-month retention period (USEPA 2012).

Europe

European Union Member states are required to introduce the requirements of the European Drinking Water Directive 98/83/EC into their national regulations. The objective of this Directive is to protect human health from the adverse effects of any contamination of water intended for human consumption by ensuring that it is wholesome and clean. It applies to all water intended for human consumption except natural mineral waters and medicinal products. Reclaimed water from municipal sewage would be required to comply with the requirements derived from Directive 98/83/EC. Individual member states can regulate additional water quality parameters or apply more stringent standards should they need to do so.

The European Commission adopted a proposal in 2018 for a revised drinking water directive to further improve the quality of drinking water. This includes provision for the addition of new parameters which may be relevant to water reuse such as chlorate, chlorite, perfluorinated compounds, and endocrine disrupting compounds (Secretary-General of the European Commission 2018).

11.5 Ion Exchange in Potable Water Reuse

Industrial water reuse applications such as boiler feedwater, make up water for power station cooling towers or ultrapure water applications typically incorporate ion exchange to produce very low total dissolved solid water. In some instances, such as the Flag Fen power station in the UK, municipal wastewater is treated to produce high purity water using membrane processes and ion exchange. The reclaimed water eases the pressure on the local potable water treatment plant which formerly supplied the power station, enabling it to deliver more of its treated water output to domestic customers.

Potable water reuse schemes are typically centered around membrane processes such as RO, and UF with ion exchange not normally featuring in the treatment stream. Ion exchange could nevertheless be considered for potable water reuse treatment in the following areas:

- Pretreatment upstream of membranes to improve membrane performance
- Post-treatment downstream of membranes to remove ionic species including disinfection byproduct precursors that remain in the membrane permeate

- Treatment of the membrane concentrate stream as part of a zero liquid discharge treatment scheme

Membrane pretreatment

Feedwater impurities may adversely impact membrane performance, resulting in loss of membrane permeability, damage to membrane materials, decreased membrane rejection of contaminants, increased operating pressure, or reduced membrane life (AWWA M62).

Ion exchange can remove scale-forming cations such as calcium, barium, silica, and strontium which can impact membrane performance (AWWA M62). This can be particularly significant when treating for other constituents such as boron that require elevated pH through the membranes.

Boron is a public health parameter with concentration guidance or limits set by organizations such as WHO, USEPA, or localities. It can be found in elevated concentrations in municipal wastewaters and exists primarily as undissociated boric acid (H_3BO_3) (WHO 2017). The relative concentrations of borate and boric acid are dependent on the pH, temperature, and salinity of the water. RO can be used to remove boron, but it is desirable to raise the pH of RO feedwater to around 10, which is above the pKa value of the soluble boric acid species. At this elevated pH, scale forming cations can cause membrane fouling problems. Pretreatment using a weak acid cation exchange system can be used to remove such scale forming foulants upstream of the RO membranes (AWWA M62).

Dissolved organic carbon (DOC) may also be removed by fluidized bed ion exchange processes such as the IXOM MIEX process. This application of ion exchange may offer the opportunity for coagulant reduction and minimize the potential for membranes organic fouling. Refer to Chapter 4 for additional information on fluidized bed ion exchange processes.

Membrane permeate polishing

The provision of multiple barriers is an important consideration in potable water reuse schemes to ensure the removal of all pathogens and constituents liked with health concerns. These include compounds which contribute to the formation of elevated and sometimes unique disinfection byproducts (WHO 2017).

Ion exchange may be used as part of a multi-barrier approach when contaminants are not sufficiently removed by other water reuse processes. Examples may include the following:

- Nitrate as discussed in Chapter 7 of this book may be present in significant concentrations in the feedwater from municipal wastewater treatment works. Ion exchange was employed in the 204 ML/d (54 mgd) Millard H. Robbins Jr. Water Reclamation Facility built by the Upper Occoquan Service Authority. However, it was only employed in standby mode to assure that total Kjeldahl nitrogen (TKN) and nitrate nitrogen discharge limits were met (Edzwald 2011).
- Ammonia (NH_3) may be poorly removed by RO membrane processes depending on the pH of the water being treated. At higher pH, the ammonium ion (NH_4^+) is converted to ammonia gas, which is

not rejected by RO. It is important to note that ammonia will not be removed by ion exchange either unless it is converted to the ammonium ion by lowering the pH.

- Bromide or iodide, which can contribute to disinfection byproduct formation following drinking water disinfection (WHO 2003).
- Boron may be removed from membrane permeate using ion exchange resin with high selectivity for boron (DuPont 2021).

11.6 Summary

While ion exchange is not typically used as part of major water reuse facilities, it can have specific applications depending on source water constituents of concern, as a pretreatment or polishing treatment to more prevalent water reuse treatment technologies such as RO. Ion exchange could play a larger role in water reuse applications for pretreatment upstream of membrane process, and in treatment streams where the selected membranes do not remove all the contaminants of concern. Ion exchange may also find niche applications as more emerging contaminants are identified and regulated to lower concentrations.

REFERENCES

DuPont. 2021. "AmberLite™ PWA10." AmberLite PWA10. January 2021. https://www.dupont.com/products/amberlitepwa10.html.

Edzwald, J.K. 2011. *Water Quality and Treatment: A Handbook on Drinking Water, Sixth Edition*. 6th ed. AWWA, American Society of Civil Engineers, McGraw-Hill. https://www.awwa.org/Store/{ProductName}/ProductDetail/{ProductId}?productId=6292.

Secretary-General of the European Commission. 2018. *Proposal for a DIRECTIVE OF THE EUROPEAN PARLIAMENT AND OF THE COUNCIL on the Quality of Water Intended for Human Consumption*. Brussels.

USEPA (US Environmental Protection Agency). 2017. "2017 Potable Reuse Compendium," 203.

USEPA. 2012. "2012 Guidelines for Water Reuse" AR-1530 (EPA/600/R-12/618): 643.

WHO (World Health Organization). 2017. "Potable Reuse: Guidance for Producing Safe Drinking-Water." *WHO*, 152.

WHO. 2003. "Boron in Drinking-Water." World Health Organization. https://www.who.int/water_sanitation_health/dwq/boron.pdf.

Wong, J.; Alspach, B.; and Chalmer, B. 2018. *M62 Membrane Applications for Water Reuse*. 1st ed. AWWA.

CHAPTER 12
Simultaneous Compliance

12.1 Introduction

The term simultaneous compliance was introduced by the US Environmental Protection Agency (USEPA) in March 2007 when it issued the Simultaneous Compliance Guidance Manual for the Long Term 2 Enhanced Surface Water Treatment Rule (LT2) and Stage 2 Disinfectants and Disinfection Byproducts (Stage 2 D/DBP) Rules. The manual states,

> "For the purposes of this guidance manual, *simultaneous compliance* means compliance with all existing Safe Drinking Water Act (SDWA) regulations....While systems may be concerned with issues pertaining to emerging contaminants, the guidance manual is not designed to address these concerns and does not discuss these issues."

The manual further states that "in addition to the SDWA regulatory issues, systems should always weigh operational and compliance with other environmental regulations when considering a treatment change" and that "readers should seek additional guidance and other technical references for addressing these compliance issues" (for example discharge permits, sludge disposal, etc.).

Although the USEPA manual was written in the context of the LT2 and Stage 2 D/DBP rules, the guiding principle is still valid for ion exchange. The implementation of any new treatment system will change the finished water quality. The regulatory and non-mandatory (secondary) water quality issues should be addressed during the design stage of an ion exhange system as well as during the operation of the ion exchange system.

This chapter describes the potential impacts of ion exchange on distribution system water quality and methods that can be implemented to quantify and address these impacts. This chapter discusses:

- the potential health and aesthetic water impacts associated with implementation of ion exchange,
- relevant regulations,
- how water the water quality impacts relate to the distribution system,
- a description of water quality parameters that can be used to assess the potential impacts associated with implementation of ion exchange,
- a description of methods that can be used to evaluate the impact of ion exchange on distribution system water quality, and
- components of a water quality monitoring plan that could be used as part of the commissioning of ion exchange systems, described in more detail in Chapter 14.

Any change to the water treatment process can change the finished water quality, and some of these changes can be unpredictable and unintended. These changes can impact the safety of the water quality or can alter the aesthetics. One of the primary concerns of treatment changes such as the installation of an ion exchange system are the impacts on corrosion. The Lead and Copper Rule Revisions (LCRR) were published in the federal register in January 2021. Changes to the rule include a 10 µg/L trigger level as well as the use of a 5th liter sample for compliance with the 15 µg/L action level for systems with lead service lines. Although the LCR compliance lead action level is 15 µg/L as of December 2020, USEPA has set the maximum contaminant level (MCL) goal for lead of zero because there is no safe level of lead in drinking water (USEPA 2016). The American Academy of Pediatrics recommends that water systems evaluate methods to reduce lead to less than 1 µg/L.

Ion exchange resins can impact the finished water by not only removing the ion of concern, but the resins may also:

- remove additional anions and cations,
- add a cation or anion,
- change the pH, and
- harbor microbes.

These changes can impact compliance with regulations. Regulations that address potential impacts associated with ion exchange are discussed below.

12.2 Relevent Regulations

The most relevant rules with regards to potential impact from implementation of ion exchange treatment include the following:

- Lead and Copper Rule (LCR)
- Stage 2 D/DBP Rule

- Revised Total Coliform Rule (RTCR)
- Filter Backwash Rule

In addition to the USEPA regulations, some state regulators have more stringent limits than the USEPA requirements and/or issued regulations for water quality parameters without a USEPA MCL:

- Perchlorate—refer to Chapter 8
- Per- and polyfluoroalkyl substances (PFAS)—refer to Chapter 10
- Chromium VI—refer to Chapter 9 (California has a regulation that is pending.)

Understanding of the mechanisms associated with lead corrosion continue to be improved as a result of research being conducted by AWWA and other industry organizations. It is no longer satisfactory to evaluate corrosion indices and achieve a value that satisfies "industry accepted" parameters. Instead, experiments used in conjunction with enhanced sampling and monitoring of the distribution system is beneficial to identify changes in corrosivity. Before pursuing an ion exchange project, the project team should coordinate with the local regulators regarding requirements with respect to simultaneous compliance concerns.

AWWA M58—Internal Corrosion Control in Water Distribution Systems provides an example of an anion exchange system implemented for nitrate and uranium removal that would reduce the pH and contribute to leaching of lead and copper into the drinking water. The anion exchange system removes bicarbonate ions, which reduces the alkalinity and lowers pH during the initial operation (approximately 400 bed volumes [BV]). Much of this effect can be mitigated by:

- blending with an untreated stream,
- fast rinse step,
- blending with vessels in staggered operation, and
- pH adjustment may be needed to avoid corrosion impacts.

In the case of single-use anion exchange resins or non-regenerable anion resins, once the resin has come to equilibrium with the water, post pH adjustment may no longer be necessary. In the case of brine regenerable anion resin systems, pH and alkalinity reduction as well as increase in the chloride sulfate molar ratio (CSMR) will be evident after every regeneration and mitigating options listed above may need to be used to control these unwanted changes.

In addition to considerations with respect to the LCRR, if there is microbial activity in the influent water, these microbiological constituents may colonize on the resin. Coliform and other bacteriological parameters such as heterotrophic plate counts (HPC) and adenosine triphosphate (ATP) should be monitored. Depending on the concentration of raw water microbial contaminants, there is the potential for resins to change the microbial characteristics of the finished water. Microbial activity can be of particular consideration with single-use resins since they are not backwashed and disinfectants are not applied to the resins. Sampling of feed and effluent water quality for microbial contaminants may be beneficial to quantify these impacts.

The Filter Backwash Recycling Rule allows treatment plants to recycle spent filter backwash to the influent of the plant. Recycle streams must be introduced at a point in the treatment plant that incorporates all treatment processes of a conventional or direct filtration system to reduce the opportunity for recycle practices to adversely affect plant performance. This is an important consideration in the design phase of the project. If ion exchange is employed at a facility that recycles spent filter backwash, the impact of the ion exchange backwash waste should be evaluated.

12.3 Distribution System Water Quality Impacts

Water quality changes need to be considered not just at the effluent of the treatment plant, but consideration should also be given to how water quality changes throughout a distribution system over time. Changes in water quality in the distribution system can occur due to:

- multiple entry points to the distribution system, which are turned on and off based on daily and seasonal demands;
- use of both groundwater and surface waters for supply of the distribution system;
- use of different treatment at each groundwater source;
- use of different treatment at each surface water source;
- use of interconnections that have different water quality from the point of entry supplies;
- changes in water quality as it ages in the distribution system. These changes occur in the piping and in the storage tanks,
- the use of aeration in water storage tanks to remove trihalomethanes; and
- the use of booster chlorination/chloramination.

In the United States, water quality sampling in the distribution system is typically based on population served. The sampling includes disinfectant residual, total coliform, disinfection byproducts, lead and copper, and associated water quality parameters listed below in Section 12.4.

Typically, utilities have a system to document and monitor customer complaints for discolored water, and taste and odor complaints. The customer complaint database should be evaluated when implementing ion exchange.

In order to assess the potential impact of ion exchange, it is helpful to understand what portion of the distribution system will be impacted by the new ion exchange system. For example, if the portion of the distribution system has no lead services, then a pipe loop-type (see Section 12.5.3) study may not be necessary for the new ion exchange system.

The development of baseline distribution system water quality can be valuable for:

- assessing the potential impacts of the new ion exchange at the feasibility level stage—for example a new anion exchange system may increase chloride levels thus elevating the Chloride to Sulfate Mass

Ratio (CSMR) and potential for lead corrosion. Additional studies would be beneficial in this case to assess potential problems and alternate anion resin conditioning to minimize impact;
- quantifying the impacts of ion exchange as part of any potential pilot testing; and
- quantifying the impacts of ion exchange upon completion of commissioning—the baseline historical data can be compared to the "post" ion exchange system.

The next section discusses which water quality parameters are important to monitor.

12.4 Impact of Ion Exchange on Water Quality Parameters

The performance of regenerable resins is repeatable as regeneration will be performed generally every day to every few weeks. Unless there is a significant change in feedwater quality, the finished water quality from the ion exchange system will be stable once it has reached equilibrium. As described in Section 2, the use of multiple vessels in parallel assists in reducing the overall change in finished water quality for regenerable resins. Single-use resins typically undergo a pre-rinse of between 50 and 1,000 BV. During this conditioning period, pH, alkalinity, and CSMR will change until steady state is reached. Alternatively, a buffered resin can be used to avoid these types of changes. Enhanced monitoring during the conditioning period is necessary to determine when the water quality is acceptable.

As described in Chapter 2: Fundamentals of Ion Exchange, a comprehensive water quality analysis of the feedwater is necessary to select the ion exchange resin. In addition to an ion analysis, the Water Quality Parameters (WQPs) from the LCR should be measured:

- pH
- Alkalinity
- Calcium
- Conductivity
- Temperature
- Orthophosphate, polyphosphate, zinc, or silica if corrosion inhibitors are used
- Silica (if a silicate-based corrosion inhibitor is applied)

Other recommended parameters to monitor may include the following:

- Chloride
- Sulfate
- Nitrate

Changes in these parameters over the course of an ion exchange run should be quantified. The ion exchange resin manufacturer can provide an estimate of the changes with computer modelling or a pilot test can be performed.

12.4.1 Corrosion WQPs

AWWA's *M58 Internal Corrosion Control of Water Distribution Systems*, 2nd edition, can be a valuable resource with respect to determining the corrosion impact of WQPs that can be altered by ion exchange. These parameters include pH, chloride, sulfate, nitrate, alkalinity, and corrosion indices.

pH: Water pH is a crucial WQP. With respect to corrosion, it is well established that lower pH water, or acidic conditions, will have a higher propensity for corrosion. Low pH water has the propensity to dissolve and dissociate metal surfaces. Maintaining a consistent pH above 7 throughout the distribution system is important for minimizing lead and copper corrosion and discolored water.

Chloride and Sulfate: CSMR is a common metric for corrosion assessment. CSMRs above about 0.6 have shown to have a corrosive effect. Chloride can be introduced to the finished water through an ion exchange system. Many resins release chloride into the water, especially during start-up. Often, the initial rinse water is discharged to waste due to elevated chloride levels at start-up (Lead and Copper Handbook) (Nguyen et al. 2010).

Water Quality Indices: Water quality indices such as the Langelier Saturation Index (LSI), Ryznar index, Aggressive Index and others can be used as a tool to identify the corrosion and scaling potential of water. The Langelier and Ryznar saturation indices are a commonly used tool to assess the potential of a water to precipitate calcium carbonate. Calcium carbonate can form a scale on pipe walls and protect the pipe from corrosion. These indices can be informative reference points in determining water corrosivity but are not conclusive.

Microbial Activity: Total coliforms and, in some cases, HPC require monitoring under TCR. Microbial activity within the distribution system can result in microbial induced corrosion, subsequent rapid pitting attack, and ultimately, pipe failure. If bacteria in the influent water colonizes on the resin, high pressure drop and the inability to satisfy the flow requirements can result. To assist in understanding the changes in microbial activity, HPC and ATP can be monitored from the effluent of the ion exchange system and within the distribution system. USEPA has indicated that an HPC of less than 500 CFU/mL is acceptable. There are no USEPA guidance values for ATP. However, utilities can often develop system specific guidance levels for ATP that are indicative of high levels of microbial activity.

12.4.2 Impacts to Disinfection By-products (DBPs)

There are potential impacts of pH on the formation of DBPs within a treatment process that could result from the implementation of ion exchange treatment. Changes in pH can create shifts in the balance of trihalomethanes (THMs) and haloacetic acids (HAAs), with lower pH values favoring the formation of HAAs and higher pH values favoring the formation of THMs (Hung et al. 2017).

12.5 Methods to Evaluate the Impact of Ion Exchange

As discussed, ion exchange can have an impact to water quality and the distribution system. Several methods have been established to help evaluate the impacts of changes to water treatment such as ion exchange. These methods include computer modeling, jar testing, and pipe loops.

12.5.1 Computer Modeling

Resin manufacturers have developed models that can predict the changes to pH, anion, and cation breakthrough. These models are especially useful where water quality is well understood.

12.5.2 Testing

Pilot or bench studies also useful to evaluate the impact of ion exchange on simultaneous compliance for conditions such as:

- the use of a new resin;
- water systems with lead service lines;
- water systems with a significant number of main breaks that could be caused by corrosion of water mains;
- water systems with a significant amount of discolored water complaints; and
- regulatory requirements—the LCRR requires that the impact of new treatment technologies evaluate the impacts on corrosion and in some instances the regulations require some form of testing.

Generally, bench level testing consists of obtaining effluent samples from the ion exchange system and testing for corrosion parameters using metal coupons or harvested piping.

Immersion or jar testing for lead and copper corrosion involves submersing lead or lead solder coupons in a jar of effluent water. There are many variations of jar testing protocols that have been developed to test corrosivity and lead leaching potential. The jars would sit for a stagnation period, then water from the jar is sampled and analyzed for lead to determine the corrosivity of the ion exchange effluent. This process can be repeated by changing the water and setting for another stagnation period over the course of several weeks (Cornwell and Wagner 2019). Immersion testing generally occurs for a minimum duration of six weeks. Figure 12-1 shows examples of a jar testing set-up.

For a regenerable ion exchange system with a run time of less than one week, bench testing can be an effective method to assess the impacts of corrosion. One can determine if pH and alkalinity adjustment is needed, what types of chemical are needed for adjustment, and the chemical dosage.

Bench or jar testing can also be used to evaluate DBP formation potential. There are several established formation potential tests including the uniform formation conditions (UFC) test and the simulated distribution system (SDS) test.

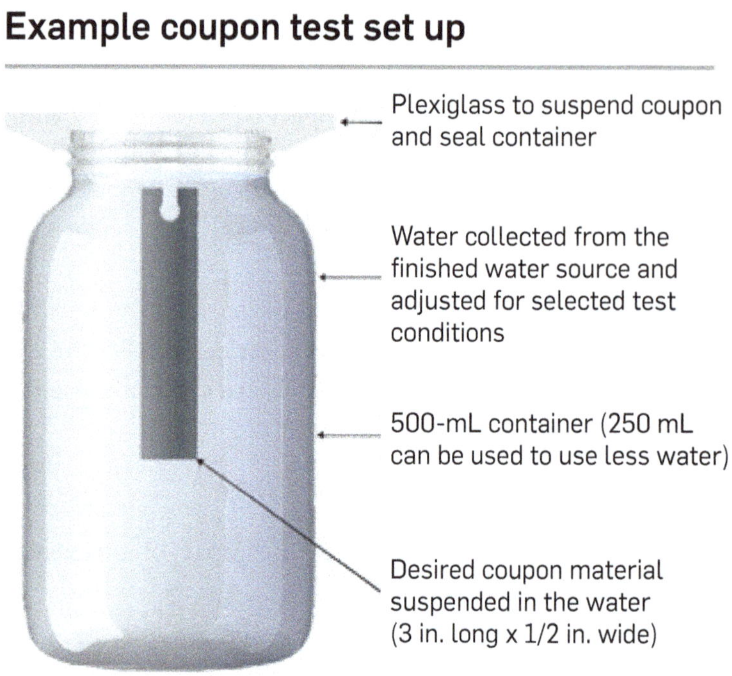

Figure 12-1 Example coupon test apparatus (Cornwell and Wagner 2019)

The idea behind these tests is to simulate effluent water disinfectant levels, and stagnation times in a jar or bottle and then test for TTHM or HAAs. The results will suggest if a DBP problem may be expected in the distribution system.

If lead and copper corrosion is a significant concern, enhanced bench or pilot-scale testing such as pipe-loop type experiments may be necessary.

12.5.3 Pipe Loop-Type Testing

Pipe loops can be used to verify immersion or jar testing results or can be used to monitor metals release in a water system before implementing a treatment change where lead and/or copper corrosion is of significant concern. Pipe loops also will take into account erosion impacts. Pipe loop type testing involves harvested or new pipe sections or metal coupons/plates where a continuous supply of treated water flows and stagnates in the pipe loop apparatus, similar to a customer's home, to simulate premise plumbing. A pipe loop can provide long-term monitoring of corrosion parameters and can be used to identify trends in metal release over time. A conditioning period is necessary for pipe loop experiments and the conditioning period can vary. For example, a Process Research Solutions (PRS) Monitoring Station, which is a flow-through pipe loop-type experiment using metal plates, can be conditioned in 1–3 months. A flow-through pipe loop containing harvested and new lead pipes may take over a year to condition. The type of pipe loop for a corrosion evaluation is selected based on cost, physical space, and duration of the conditioning period. In order to use a pipe loop-type test to evaluate the impact of an ion exchange system on corrosion, an ion exchange pilot system is necessary. The pipe loop would receive the effluent from the ion exchange system. The water from the pipe loop would normally go to waste. Examples of pipe-loop type testing are shown below.

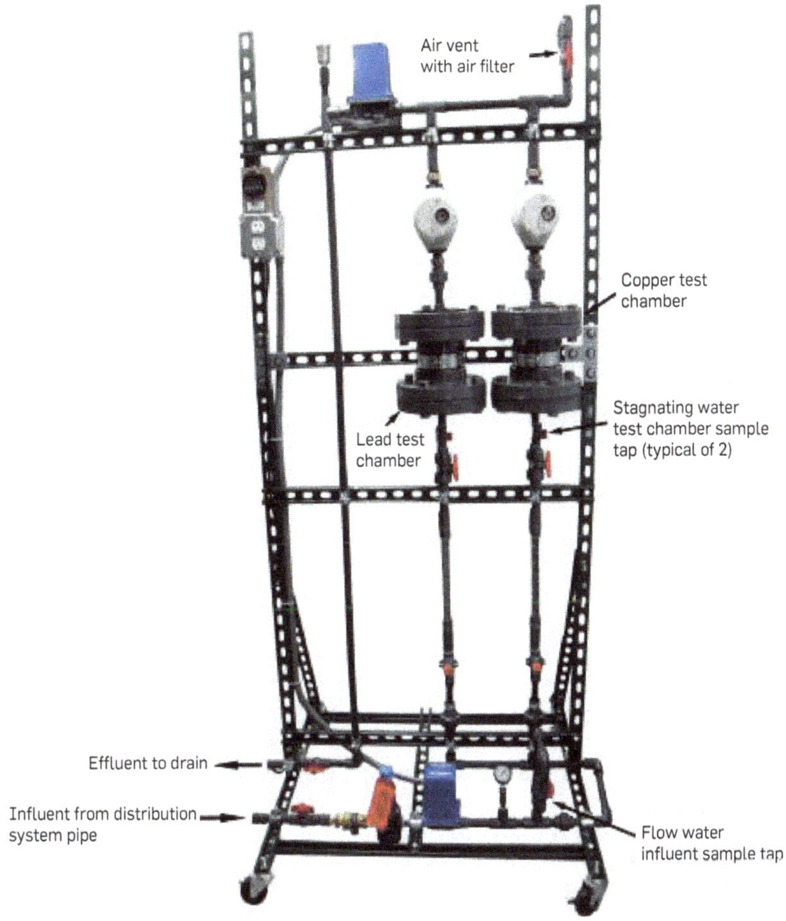

Figure 12-2 PRS monitoring station setup (4-lead/copper chambers)

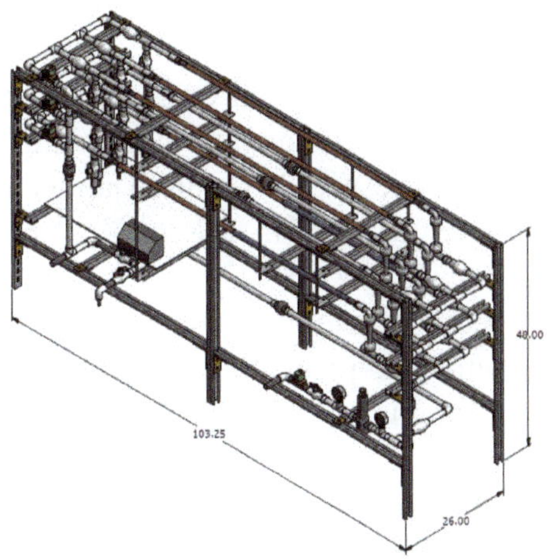

Figure 12-3 Flow-through pipe loop with nine test pipes including harvested and new lead, and 1/2 lead–1/2 copper to simulate a partial replacement

For further information on treatment change impacts and monitoring, please reference AWWA's *M56 Internal Corrosion Control in Water Distribution Systems.*

Chapter 13 describes ion exchange pilot testing.

12.6 Components of a Water Quality Monitoring Plan

The LCR requires utilities to monitor certain WQPs in order to assess a system's vulnerability to corrosion. Monitoring is performed at the Entry Point to the Distribution System (EPTDS) and sampling points in the Distribution System (DS) and also includes parameters that indicate the presence of corrosion inhibitors such as ortho and polyphosphate, zinc or silica, if applied. Required parameters to monitor include the following:

- pH
- Alkalinity
- Calcium
- Conductivity
- Temperature
- Orthophosphate (if a phosphate-based corrosion inhibitor is applied)
- Silica (if a silicate-based corrosion inhibitor is applied)

Other recommended parameters to monitor may include the following:

- Iron
- Manganese
- Aluminum
- Chloride
- Sulfate
- Nitrate

To understand the impacts of ion exchange on distribution system water quality, a distribution system water quality monitoring plan should be developed. This plan is distinct from the WQPs required under the LCR. In developing a distribution system water quality plan, the utility should review their regulatory and nonregulatory sampling programs.

The first step towards developing a viable WQP sampling plan is to contact your local state environmental protection agency or health department to determine if there are specific guidance and reporting requirements for your water treatment facility. This may include templates for developing a WQP sampling plan or reporting templates. As mentioned in Section 12.3, WQP sampling associated with the LCR is usually conducted during each six-month lead and copper monitoring period. However, individual states may require additional or more frequent sampling. The components of a WQP sampling plan generally include the following items:

- **General Information about the Water System**—this could include but may not be limited to the following:
  - Water system name and PWSID (public water system identification)
  - System type (CWS [community water system], etc.)
  - Water system source(s)
  - List of treatment facilities and interconnections
  - Population served and number of connections
  - Distribution system map
  - Other pertinent information about the water system
  - Contact information

- **Sampling Sites**—The number of required LCR WQP distribution sampling sites are generally determined by the water system size, or population served and can range anywhere from 1 to 10, or as specified by your state regulator. A suitable pool of sampling sites should be developed throughout your distribution system. Consideration should be given to the following:
  - Areas of particular concern with regards to corrosion control
  - Areas where different sources are present
  - Areas where booster chlorination or other treatment is employed
  - Areas of high water age or a change in pressure district
  - Sensitive populations.

Development and changes to the pool should be made with the collaboration of your state regulator.

- **Monitoring Schedules**—WQP sampling requirements and timing can vary depending on a number of factors including whether or not the utility has installed corrosion control treatment (CCT) if source changes occur, or if interconnections are operating. A utility will need to work closely with their state regulator to determine appropriate sampling protocols for initial and follow-up monitoring programs.

- **Sample Collection**—your sampling plan must provide details on the proper collection of samples by either a certified laboratory or other approved party, including the removal of faucet aerators, proper flushing times, observations on the clarity or color of the water, proper storage and shipping of the samples, etc. The contact information of the sample collector(s) should be provided.

- **Sample Analysis and Reporting**—Analysis should be performed by a certified laboratory, with the exception of temperature and pH which must be performed by a certified analyzer in the field. Some utilities employ certified personnel that have the credentials to properly collect and analyze samples, including calibration of equipment, proper record keeping, execution of quality assurance/quality control (QA/QC) tasks, etc. Reporting of test results are generally performed by the certified laboratory. Utilities must follow state guidelines and use state reporting forms (if applicable) when reporting all WQP sample results.

- **Compliance Requirements (Sampling and Reporting)**—the WQP sampling plan should also provide details on the compliance requirements associated with collection and reporting of WQPs. This includes the proper timing of sample collection, reporting requirements, and the penalties and violations associated with failure to collect, analyze, and report WQPs in accordance with the developed sampling plan. State requirements may vary, but general concerns include; failure to collect appropriate samples within the designated time period, improper analysis or QA/QC practices, and failure to report WQP results as instructed by the state regulator.
- **Compliance Requirements (Treatment Technique)**—this applies to utilities that have optimized corrosion control treatment (OCCT) in place and have developed target levels or ranges for the parameters associated with this treatment. For example, limits for pH or orthophosphate levels have been developed and have been approved by the state. Excursion of these target levels or ranges can result in a treatment technique (TT) violation which may require development of a specific action plan to address the excursions, public notification, redevelopment of the OCCT treatment. Utilities must work closely with their state regulators to ensure that the WQP Sampling Plan accurately describes the actions required if an excursion of a WQP result occur.

A WQP sampling plan is a tool to assess and monitor potential changes due to the addition of an ion exchange system, but it is not comprehensive. In addition to the monitoring described above, monitoring should include routine sampling for biological concerns, such as coliform testing, and for DBP formation in the system. Simultaneous compliance means that water quality needs to meet the requirements of all regulations. Required LCR monitoring should be combined with Revised Total Coliform Rule (RTCR) and DBP monitoring. If monitoring indicates that there are, or could be concerns with the water quality, adjustments may need to be made to the ion exchange system or other components of the treatment system.

While monitoring after installation is important, good baseline (pre-ion exchange) data should first be established. Without such data, it could be difficult to determine if any concerns about water quality are due to the ion exchange system or were pre-existing. Utilities should begin regular monitoring of the parameters described above for at least six months prior to ion exchange installation, if they do not already do so.

REFERENCES

Cornwell, D.A. and Wagner, J.R. 2019. "Coupon Procedures for Evaluating Lead and Copper Solubility." *Journal AWWA,* 111 (10).

Hung, Y.-C.; Waters, B.W.; Yemmireddy, V.K.; and Huang, C.-H. 2017. "PH Effect on the Formation of THM and HAA Disinfection Byproducts and Potential Control Strategies for Food Processing." *Journal of Integrative Agriculture* 16 (12): 2914–23. https://doi.org/10.1016/S2095-3119(17)61798-2.

Nguyen, C.; Stone, K.; Clark, B.; and Edwards, M. 2010. "Impact of Chloride:-Sulfate Mass Ratio (CSMR) Changes on Lead Leaching in Potable Water." WRF Project #4088. Denver, CO: Water Research Foundation.

USEPA (US Environmental Protection Agency), OW (Office of Water). 2016. "Basic Information about Lead in Drinking Water." Overviews and Factsheets. USEPA. February 2, 2016. https://www.epa.gov/ground-water-and-drinking-water/basic-information-about-lead-drinking-water.

CHAPTER 13
Pilot Testing

13.1 Introduction

Many ion exchange systems can be designed and commissioned based on an assessment of existing baseline raw water quality data, as described in Chapter 2, and by establishing well-defined finished water quality goals. In many instances, there is sufficient empirical data on resin performance to determine the number of bed volumes that can be treated before regeneration or resin replacement is needed. Chapters 3 through 10 present ion exchange system design examples for several common contaminants found in water sources used for drinking water supply. A successful ion exchange system will not only meet the water quality objectives, but will also consider additional criteria such as:

- construction cost objectives,
- energy use and cost,
- chemical usage and cost,
- waste production and cost,
- maintenance requirements, and
- regulatory requirements.

Because source water quality can be highly variable, existing empirical information may not be fully predictive of full-scale performance. It is prudent to conduct testing in the form of bench or pilot testing to assess each criterion and demonstrate effectiveness. These trials can be useful in refining a preliminary design and comparing the performance of alternate resins. In some situations, comparison with a similar facility that has a full scale ion exchange system or that performed ion exchange testing may also be used in place of bench or pilot testing.

Reasons for conducting pilot testing include the following:

- Test new process
- Predict process performance
- Document process performance
- Optimize system design
- Satisfy regulatory agency requirements
- Satisfy legal requirements
- Verify performance claims of the manufacturer

(Crittenden et al. 2012)

This chapter provides an overview of how bench, pilot and demonstration testing can be implemented for ion exchange projects. The chapter includes guidance on the following:

- Differences between bench, pilot and demonstration testing
- Development of a comprehensive and detailed protocol
- Coordination with regulators
- Establishing the objectives of the pilot plant testing
- Configuration of the pilot plant including the resin columns and ancillary equipment
- Residuals handling
- Water quality sampling
- Data analysis
- Schedule
- Development of a pilot report

13.2 Bench, Pilot, and Demonstration Testing

13.2.1 Bench Scale Testing

Bench scale testing, sometimes called lab testing, is a short duration test generally lasting a few hours to a few days to determine the effectiveness of a specific resin or resin technology. In many cases, a resin supplier will perform bench testing for a customer or prospective customer. The actual source water should be used for the testing so that water quality parameters are representative of actual conditions. Water for bench testing can be shipped to the supplier to perform the test or the supplier may also provide a test kit to the owner so that the owner or an owner representative, such as an engineering firm, can perform the test. The results of the bench scale test will provide information to assess if the specific resin or resin technology can meet the water quality goals as well as important operational parameters, such as an estimate of volume that can be treated before regeneration.

Benefits of bench scale testing include the fact that it is quick and has a low cost. Limitations of bench scale testing include: the water sample represents a single-water quality event; only a small data set is generated; and the study does not replicate the full-scale system as it does not contain the regeneration components—regenerated resin will have less capacity than fresh resin.

Bench scale testing is particularly well suited to compare the effectiveness of different resins to each other. Since it does not include the same type of configurations as a full-scale unit, it may not represent the exact water quality results as would be seen full scale.

13.2.2 Pilot Testing

Pilot testing is a smaller scale version of the full-scale plant and contains all of the components of the full-scale plant, including regeneration where applicable and residuals handling systems. The pilot plant generally operates at flow rates between 0.5 and 10 gpm. Pilot plants are typically operated for several weeks to several months in order to assess resin performance under a range of source water quality conditions and allow optimization of the design criteria to achieve the cost and residuals objectives. Pilot testing may also be required by the regulatory agency, especially if a proposed resin has never been used in that state and/or local regulatory jurisdiction. Pilot testing costs can range from several thousand to several hundred thousand dollars. The major cost components of ion exchange pilot testing consist of the following:

- Pretreatment equipment that may be needed to remove contaminants such as turbidity, iron, and manganese and those described in the previous chapter
- Resin columns and regeneration equipment where applicable
- Waste handling
- Analytical testing
- Labor

Given these costs, pilot testing is generally suitable when the new treatment plant capacity is at least 100 gpm. When a treatment plant capacity is below 100 gpm, it may be more cost-effective to purchase a full-scale system and perform a demonstration study.

Benefits of pilot scale testing include: more data that accounts for variation in source water, results are more representative of expected full-scale performance, and it allows for waste characterization and optimization of design parameters.

Figure 13-1 provides an example of the components and configuration of trailer mounted pilot plant for a fluidized bed ion exchange system.

A pilot will require regular maintenance to ensure the flows are steady and the prefilters are changed out when necessary. Data collected from the pilot is analyzed to determine which of the tested media performs best.

Figure 13-2 is a photograph of a ion exchange pilot skid. Columns, pressure gauges, and valves can easily be seen.

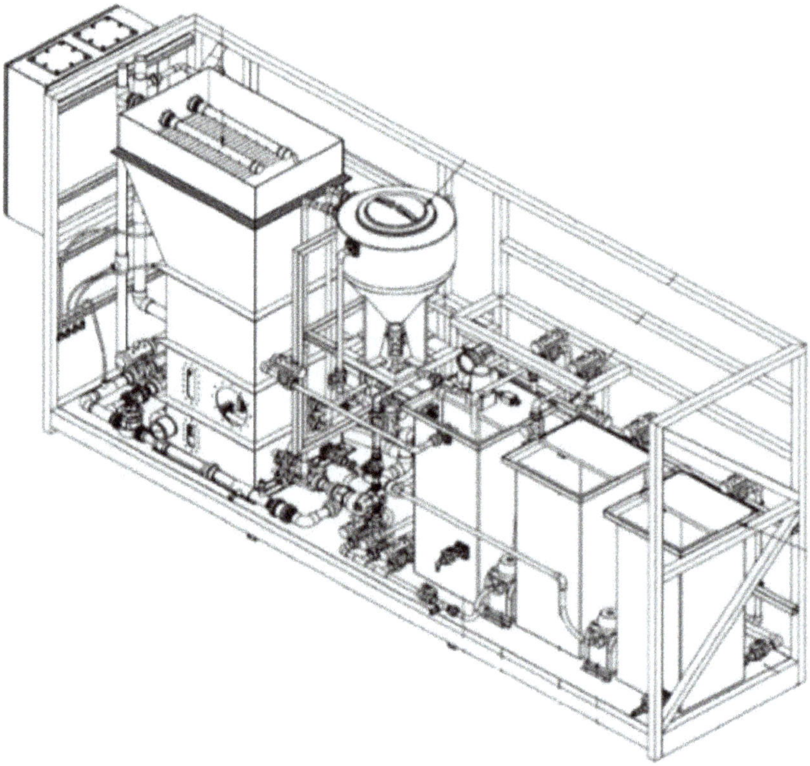

Figure 13-1 Example fluidized bed 10 GPM pilot plant (MIEX)

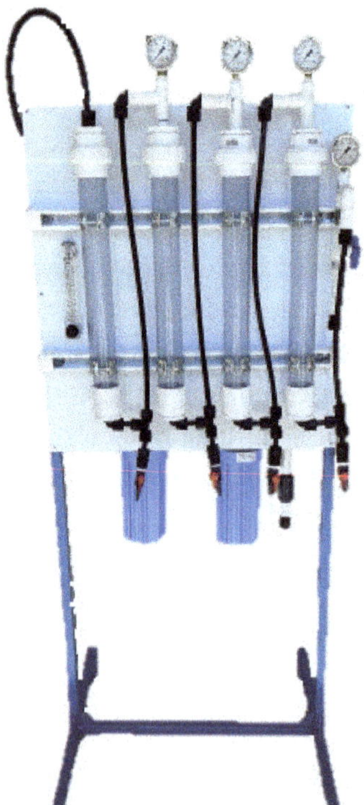

Figure 13-2 Pilot skid before media addition
Courtesy of Purolite Corporation

13.2.3 Demonstration Testing

Demonstration testing is the use of temporary equipment that could be converted to permanent equipment to treat the entire source of supply. Depending on the requirements of the regulatory agency, it may be possible to send the treated effluent to the distribution system. Regulators may require a temporary permit and additional testing to ensure the treatment is meeting the treatment targets. The demonstration testing system should include disinfection and corrosion control systems. The demonstration test mitigates potential scale up issues associated with pilot plant systems. Generally, a demonstration test will involve collecting more data than a pilot test to assess performance. Demonstration testing can be economical for wells with a capacity generally less than 100 gpm as this will result in a 4 ft diameter vessel operating at 8 gpm/ft^2.

13.3 Objectives of Pilot Plant Testing

Bench, pilot and demonstration testing each have different objectives to assist in determining the feasibility of a full-scale permanent ion exchange system. Generally, the purpose of a bench test is a quick go or no-go determination. It can be used to screen different media to determine whether ion exchange is feasible for treatment, or which specific resins are likely to be most effective at achieving the treatment goals and/or cost objectives. Bench testing can be performed prior to pilot testing to measure the effectiveness of ion exchange to solve a specific water quality issue.

There are often multiple objectives of a pilot test. One objective is to validate the theoretical modeling so that both the water agency and the regulators know that performance can be predicted. Another goal is to assess resin performance under a range of water quality conditions. Finally, pilot testing can be used to optimize the design to reduce costs and residuals.

The objectives of pilot testing may include the following:

- Comparing the performance of multiple resins (at least two if possible)
- Determine the number of bed volumes that can be treated before breakthrough
- Evaluate the impact of multiple surface loading rates
- Evaluate the impact of multiple empty bed contact times
- Quantify chromatographic peaking of contaminants that may result in higher concentrations in the effluent than their raw water concentrations and may cause an exceedance of the regulatory level or could impact the corrosivity of the effluent
- Optimize regeneration chemical requirements
- Quantify the residuals production
- Evaluate impact of ion exchange treatment on downstream treatment processes and required chemical dosing
- Validate operational costs

13.4 Protocol

A comprehensive protocol is a critical aspect of the testing program. A written protocol should be prepared for bench, pilot, and demonstration testing. The protocol is a detailed workplan that presents the objectives, scope, testing schedule and budget. The protocol should be updated whenever changes are made to the testing program. The protocol should be reviewed and compared to the actual work being performed throughout the study. The protocol can be used by the following stakeholders:

- **Regulator**—ultimately, the regulatory agency will need to approve the design criteria obtained as part of the protocol. Some projects require approval of the protocol by the regulatory agency and even for those projects that do not require formal approval, utilities are encouraged to submit the protocol to the regulators and obtain their comments prior to commencing pilot testing. This is important as the information from the pilot test will be used to develop the design documents that eventually will require regulatory approval.
- **Executives of the utility that will be funding the project**—the protocol needs to clearly demonstrate the value of the testing and how funding of the testing will result in reduced risk and improve quality. Executives may be interested in: (1) testing multiple resins to provide competitive pricing, (2) evaluation of regeneration methods to reduce chemical costs, and (3) brine waste minimization where applicable.
- **Engineers of the utility**—most utilities have an engineering group that will manage the design engineers. One of their duties is to satisfy the cost objectives of the executives and to satisfy the water quality and maintenance objectives of the operators. The pilot testing experiments will allow testing of alternate resin configurations, loading rates and regeneration scenarios to satisfy these objectives.
- **Operators at the utility**—this group will be responsible for operation of the full-scale system. This group generally prefers a system that can: (1) reliably respond to changes in water quality, (2) reliably achieve the finished water quality objectives with minimal manual operational changes, and (3) provide ease of operation and maintenance. Operators of the utility are encouraged to participate in the operation of the pilot plant and provide feedback to the design team. This feedback can assist the design team and equipment supplier in customizing the full-scale system.
- **Pilot plant operator**—although the pilot plant may have automation, an operator is needed to check on the condition of the pilot plant, take samples and make operational adjustments as described in the protocol. The duties of the operator must be clearly described in the protocol.

- **Design engineers**—the data obtained during the pilot will provide the basis for the resin and equipment selection and layout. The protocol should describe the testing scenarios that will support the design criteria.
- **Ion exchange resin supplier**—typically, the resin supplier does not provide the vessels and associated equipment. The equipment is generally supplied by an Original Equipment Manufacturer (OEM). The resins to be tested must be clearly identified. Coordination with the OEM is necessary as different resins may require different components related to backwashing and regeneration.
- **Ion exchange Original Equipment Manufacturer (OEM)**—this group provides the ion exchange equipment. In some instances, the OEM may not have a direct relationship with the resin supplier(s) and the engineer or owner will need to coordinate with the resin supplier. The project team must clearly identify existing contractual relationships and determine the requirements of data sharing with the resin suppliers and OEM.
- **Contractor installing the system**—the pilot system will require interconnecting piping and wiring and must be integrated with the existing facility. In some instances, the owner's staff can perform the installation but many times, a contractor is required. Drawings should be prepared to demonstrate the scope of work to be performed by the contractor. In some instances, there may be insufficient space for installation of the pilot equipment and a temporary structure may be needed. Temperature can affect the performance of ion exchange resins, and therefore, heat tracing or heating may be required.

Raw water quality samples should be obtained prior to start-up of the pilot system. The raw water quality should be reviewed by the resin suppliers to provide guidance on pilot configuration and pilot testing goals. When submitting the data to the resin suppliers, the utility or engineer should notify the resin supplier of any variations in the raw water quality data. The resin supplier working with the OEM can then provide a description of the pilot equipment and the engineer and owner can then determine if there is sufficient space, power, and sewer available for the pilot system.

Multiple iterations of the pilot configuration may be necessary to accommodate existing site limitations. The owner will generally obtain pricing for rental or purchase of the pilot equipment. With this information, the engineer can then prepare drawings describing the scope of work for the installation contractor and the owner can obtain pricing for installation. Upon completion of the installation, the resin supplier/OEM will start-up the equipment and the system will be turned over to the owner for operation. With the advancement in programable logic controllers (PLC) and remote data transfer technology, a pilot plant may be equipped with remote access so that the system can be monitored off site.

13.5 Pilot Plant Configuration

Pilot testing should simulate an ion exchange system that can be scaled up to the anticipated full-scale system. The pilot plant effluent has a typical flow rate between 1 and 10 gpm. The pilot plant should use the largest column possible so that column wall forces are negligible, typically 4–6 in. diameter, but in some case a column with a diameter of 2 in. may be acceptable. For regenerable pilot systems, accommodations must be made to collect and dispose of brine waste.

Pilot testing of a non-regenerable media, is easier than a regenerable media as the system doesn't require a sanitary sewer to receive the regenerant waste solution.

Figure 13-3 presents a pilot system schematic. The pilot plant will need a source of supply and, in many instances, the raw water source is not operated continuously. Continuous operation of the pilot plant is preferred. In those cases where there is not a continuous source of supply to the pilot, a temporary equalization tank will be needed. Coordination of filling of the equalization tank with the owner's operators will be required. It is essential that controls and spill containment be provided to prevent and/or contain an overflow of the equalization tank. An equalization tank does have the potential to change the water chemistry–the pH will change as CO_2 is released, iron may fall out of solution, or bacteria may start to grow. If an equalization tank is used, the water should be turned over as often as possible.

13.6 Wastewater and Residuals Handling

Wastes generated from the pilot plant include:

- treated effluent from the pilot columns,
- drain waste from any instruments,
- regeneration waste wastes, and
- spent ion exchange media especially for non-regenerable resins.

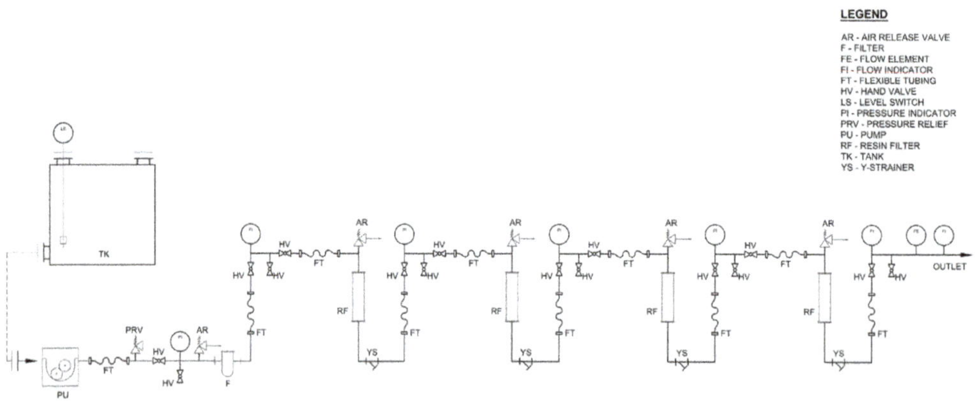

Figure 13-3 Ion exchange pilot system with equalization tank

The treated effluent of the pilot plant may not be suitable for disposal to the ground or to the local water bodies depending on local regulations. If a sewer is not available, a holding tank will be necessary. The waste from the holding tank should be trucked away for off-site disposal. This waste tank will also be used for the regeneration waste.

The spent resin will also be another waste product. For single-use resins, the waste should be characterized to determine disposal options.

When performing any radionuclide removal, check with the applicable state regulatory agency as a state specific Radioactive Materials License (RML) may be required to perform an on-site pilot test. Also, the test media used should be disposed of in a licensed or appropriate manner per regulatory requirements.

13.7 Water Quality Sampling

Water quality analyses can be an expensive component of the pilot testing. The water quality testing will consist of both field and laboratory testing. A detailed schedule should be performed that describes the water quality parameters, location of sample collection, frequency of testing, and testing method. A cost estimate of the testing should be prepared as part of the development of the protocol so the test is budgeted appropriately. Quality control of the field-testing equipment should be performed prior to sampling. This includes equipment preparation such as calibration of field analyzers, labelling of bottles, and preservation of bottles. Field blanks and duplicates are also suggested for additional quality control. A chain of custody form and field notebook should be implemented by the sampler for all water quality analyses.

Ion exchange will not only remove regulated contaminants but will also remove additional cations (for CIX) and anions (for AIX). This removal will impact the stability and corrosivity of the finished water. The stability and corrosivity will also be impacted by the ion which is "added" as part of the exchange process. Chapter 12 provides information related to corrosivity. During the pilot testing, sufficient water quality analyses should be performed to determine the impact of corrosivity.

Table 13-1 provides an example of the initial raw water quality testing that was conducted prior to installation of an ion exchange pilot system for per and polyfluoroalkyl substances (PFAS) removal.

Water quality testing should be customized based on the contaminant being removed, regulatory requirements, and funding. Table 13-2 is an example of a testing schedule for a PFAS pilot test. The influent water quality should be monitored to note any changes. The pilot effluent should be monitored for ion exchange performance. In this table the ion exchange resin was built with four segments operated in series to represent different empty bed contact times.

Table 13-1 Initial raw water quality testing

Description	Field or Lab	Lab Method	Reporting Limit	Frequency/ No. of Samples
Temp	Field	SM 2550 B	N/A	1
pH	Field	USEPA 150.1	N/A	1
Turbidity	Field	USEPA 180.1	1 NTU	1
Silica	Lab	USEPA 200.8	0.1 mg/L	1
Fluoride	Lab	USEPA 300.0	0.5 mg/L	1
Sulfate	Lab	USEPA 300.0	5.0 mg/L	1
Iron (total and dissolved)	Lab	USEPA 200.7	0.02 mg/L	1
Manganese (total and dissolved)	Lab	USEPA 200.8	0.02 mg/L	1
Magnesium	Lab	USEPA 200.7	0.1 mg/L	1
Total dissolved solids (TDS)	Lab	SM 2540 C	25 mg/L	1
Total suspended solids (TSS)	Lab	SM 2540 D	25 mg/L	1
Total organic carbon (TOC)	Lab	SM 5310 C	0.5 mg/L	1
Dissolved organic carbon (DOC)	Lab	SM 5310 C	0.5 mg/L	1
Ammonia (NH_3)	Lab	SM 4500-NH3 D	0.1 mg/L	1
Nitrite	Lab	USEPA 353.2	0.01 mg/L	1
Nitrate	Lab	USEPA 353.2	0.1 mg/L	1
Arsenic	Lab	USEPA 200.8	1.0 μg/L	1
Alkalinity	Lab	SM 2320 B	1 mg/L as $CaCO_3$	1
Calcium	Lab	USEPA 200.7	0.1 mg/L	1
Chloride	Lab	USEPA 300.0	2.0 mg/L	1
Bromide	Lab	USEPA 300.0	10 μg/L	1
Conductivity	Lab	SM 2510 B	2.0 uhos/cm	1
Hardness	Lab	SM 2340 B	0.66 mg/L as $CaCO_3$	1
Aluminum	Lab	USEPA 200.8	2.0 μg/L	1
Uranium	Lab	USEPA 200.8	1.0 μg/L	1
Perchlorate	Lab	USEPA 331.0	0.05 μg/L	1
Total chromium	Lab	USEPA 200.8	0.9 μg/L	1
Hexavalent chromium	Lab	USEPA 218.6	0.02 μg/L	1

(continued)

Table 13-1 Initial raw water quality testing (*continued*)

Description	Field or Lab	Lab Method	Reporting Limit	Frequency/ No. of Samples
VOCs	Lab	USEPA 524.2	0.5 ug/L	1
1,4 Dioxane	Lab	USEPA 522	0.07 ug/L	1
Gross alpha	Lab	SM 7110 B	3.0 pCi/L	1
Radium 226	Lab	SM 7500-Ra B	1 pCi/L	1
Radium 228	Lab	SM 7500-Ra D	1 pCi/L	1
Methylene blue active substances (surfactants)	Lab	SM 5540 C	0.1 mg/L	1
PFAS – 18	Lab	USEPA 537.1	2.0 ng/L	1

Table 13-2 Effluent water quality testing

Description	Sample Column				Field or Lab	Lab Method	Reporting Limit	Frequency	Monthly No. of Samples
	A	B	C	D					
Temp	X	X	X	X	Field	SM 2550 B	N/A	Daily	~20
pH	X	X	X	X	Field	USEPA 150.1	N/A	Daily	~20
Total organic carbon (TOC)	X	X	X	X	Lab	SM 5310 C	0.5 mg/L	two weeks*	2
Alkalinity	X	X	X	X	Lab	SM 2320 B	1 mg/L as $CaCO_3$	two weeks*	2
Chloride	X	X	X	X	Lab	USEPA 300.0	2.0 mg/L	two weeks*	2
Sulfate	X	X	X	X	Lab	USEPA 300.0	5.0 mg/L	two weeks*	2
Nitrate	X	X	X	X	Lab	USEPA 353.2	0.1 mg/L	two weeks*	2
Arsenic	X	X	X	X	Lab	USEPA 200.8	1.0 µg/L	two weeks	2
Iron (total and dissolved)	X	X	X	X	Lab	USEPA 200.7	0.02 mg/L	two weeks	2
Manganese (total and dissolved)	X	X	X	X	Lab	USEPA 200.8	2.0 µg/L	two weeks	2

(continued)

Table 13-2 Effluent water quality testing (*continued*)

Description	Sample Column A	B	C	D	Field or Lab	Lab Method	Reporting Limit	Frequency	Monthly No. of Samples
Total dissolved solids (TDS)	X	X	X	X	Lab	SM 2540 C	25 mg/L	two weeks	2
PFAS – 18	X	X	X	X	Lab	USEPA 537.1	2 ng/L	two weeks†	6

* These parameters are monitored daily for days 1 through 4 to capture the estimated period before background equilibrium was achieved and then every two weeks thereafter.

† These parameters are monitored weekly for the first four weeks, and then every two weeks thereafter.

Note: X indicates a sampling event.

Even though the pilot plant may be equipped with a data logger/PLC that will record information such as flow rates and conductivity, the pilot plant operator should manually record this information at least on a weekly basis.

Many utilities do not have the capabilities to perform the necessary laboratory services. In cases where an outside laboratory is needed, the turnaround time for results may be several weeks.

13.8 Data Analysis

Ahead of the execution of the pilot, a detailed data analysis methodology should be developed as part of the pilot protocol. This methodology will enable the project team to determine if the pilot testing achieved the objectives and to optimize the design of the full-scale system. The data should be updated at least weekly and should be reviewed with the project team no less than monthly. For ion exchange systems, as a minimum, the protocol should include the following:

- How each of the water quality parameters identified in the sample schedule will be analyzed (which parameters will be graphed together, etc.)
    - It may be beneficial to use surrogate parameters for test that have longer lead times, such as UVA254 for DOC.
- How each of the operating characteristics such as bed volume rate, surface loading rate, and regeneration methods will be analyzed
- Development of breakthrough curves for each contaminant (The breakthrough curves should be based on bed volumes and bed volumes can be converted to volumetric production from the feed sources). Figure 13-4 presents an example breakthrough curve.

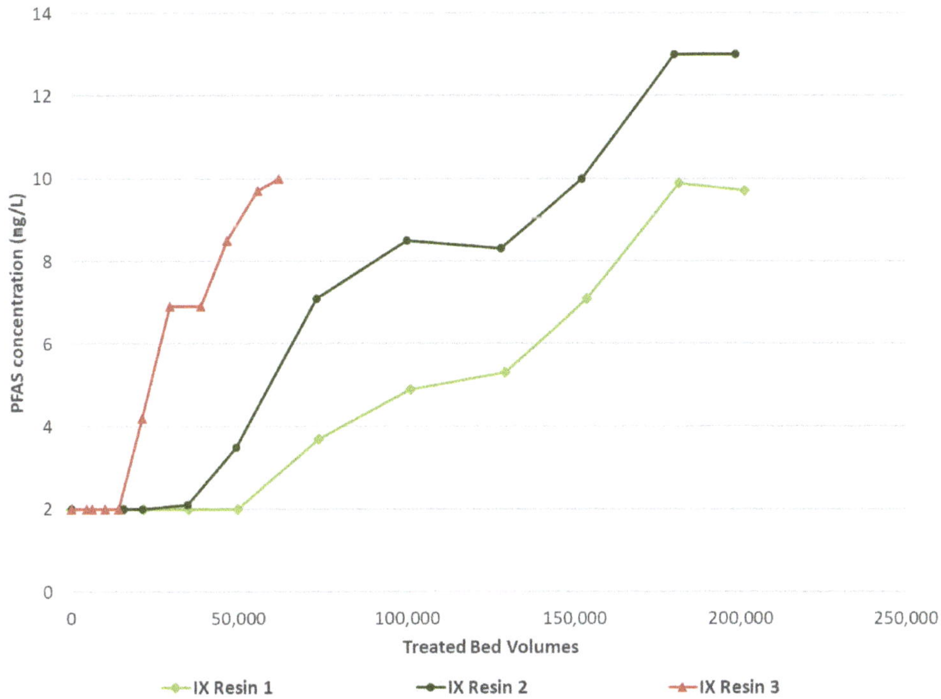

Figure 13-4 Ion exchange resin breakthrough curve of PFxHS

- Breakthrough curves should indicate the operating conditions including surface loading rate
- Definition of how the data will be stored electronically (spreadsheet, database)
- Definition of data analysis templates such as excel graphs, or other graphing and analysis programs (R, PowerBI, etc.)

There is sometimes a delay of several weeks between the time of sampling to delivery of the analytical results. This data analysis methodology should reflect the time associated with the laboratory analyses as it can impact the operation of the pilot plant and the interpretation of the results.

The pilot test is an experiment and the initial protocol should anticipate that changes will be needed to the operating conditions based on the water quality results. The development of the data templates as part of the initial protocol will expedite review of the data and facilitate operational modifications.

13.9 Schedule

The pilot protocol should contain an endpoint determination to make design decisions, satisfy the requirements of the regulators and enable an assessment of the cost of the pilot testing. The following factors will impact the project schedule:

- Source water: groundwater sources generally have much less variability than surface water sources. If the project has multiple wells, depending on their water quality variability it may be necessary to assess pilot operation with each of the wells or combination of wells. Surface water sources will vary during precipitation events and will change seasonally. The pilot testing should capture the variation in the surface water quality. For example, the fluidized bed ion exchange system is typically used for the treatment of raw surface water and the pilot testing should be of sufficient duration to measure performance under the range of feedwater quality.
- Regenerable resins: The regeneration frequency varies greatly depending on the target contaminant and the water quality. For example, nitrate ion exchange resins are regenerated every 6–48h on average. A nitrate pilot operating on a stable groundwater may operate for four weeks to allow for multiple loading rates and regeneration scenarios. Regenerable chromium VI resins may not need regeneration for two to six weeks, so in order to achieve multiple regenerations, the pilot duration may be for several months.
- Non-regenerable or single-use resins: may last for months or years in the field. In these cases, accelerated pilots may be designed. An accelerated pilot is done with the resin supplier. The resin supplier will model performance with a shortened empty bed contact time, so that results may be available in a fraction of the time. The pilot will validate the model, and provide results in a timely fashion to make equipment and treatment decisions.

13.10 Pilot Report

The report should be written to address the interests of each of the stakeholders described earlier in this chapter. Many regulators will provide input as to the contents and level of detail to be included in the report. Coordination with the resin suppliers is essential. Data associated with each resin supplier should not be shared with the other supplier unless permission is received from the supplier or from the OEM providing the pilot system. The report should provide recommendations regarding final design criteria such that the ion exchange system satisfies the water quality objectives, is cost effective, minimizes residuals generation, and addresses ease of operation and maintenance.

13.11 Coordination With Regulators

The requirements of the regulators will vary in different regions. Furthermore, the addition of ion exchange systems can have unintended consequences to the finished water, see Chapter 12: Simultaneous Compliance for additional discussion. The regulators will be responsible for approving the design documents and, therefore, regulator input is encouraged prior to pilot startup and upon completion of the pilot study. Regulatory input at the pilot stage is generally from the state or country agency responsible for drinking water. However, resins generate waste. Regenerable resins generate a brine waste. The brine waste stream must be coordinated with the appropriate authority—it may be beneficial to engage the appropriate wastewater regulators at the pilot phase. Non-regenerable resins will produce a resin waste that will require disposal. It may be beneficial to coordinate with the solid waste regulators regarding waste disposal of the contaminated resin. Engagement at this stage will likely facilitate the regulatory approvals of the full-scale project.

When performing any radionuclide removal, check with the applicable state regulatory agency as a state specific RML may be required to perform an on-site pilot test.

CHAPTER 14

Instrumentation, Commissioning, and Operations

14.1 Introduction

In Herbert Hoover's essay "Engineering," he writes "If his works do not work, he is dammed." The "works" in this statement refers to the physical facilities such as a *new ion exchange* system and the "work" refers to the *performance* of the ion exchange facility. This chapter presents an overview of the activities involved in obtaining a "working" ion exchange facility. An ion exchange facility must demonstrate it will work in the short term and demonstrate long-term reliability.

An important aspect of determining whether an ion exchange facility achieves the performance objectives is the use of measurements. These measurements include the following:

- Hydraulic measurements such as flow, pressure, and level
- Water quality measurements of the feedwater, effluent water
- Water quality measurements associated with pretreatment components and residuals handling systems
- Ion exchange resin condition through resin analysis
- Quality, purity, and concentration of regenerant chemicals
- Measurements of equipment performance such as motor overloads, motor speed, and valve position

These measurements provide data to objectively assess whether the ion exchange system is meeting the performance requirements. The hydraulic and equipment performance measurements are common to all types of process equipment and are not discussed in this chapter. An overview of water quality instruments that are used in ion exchange systems are described in this chapter.

A water treatment plant that uses ion exchange includes components provided by multiple suppliers and installed by different contractors/subcontractors. Once the system is assembled it must demonstrate that the system "works." This initial demonstration phase is often referred to as the "commissioning" phase and culminates with a "performance test". This chapter provides an overview of the major components associated with the commissioning phase including the basic elements of the performance test. Note that each individual piece of equipment in an ion exchange system will have its own operations and maintenance (O&M) manual and the treatment plant as a whole will have an O&M manual. This chapter is not a substitute for robust O&M manuals.

The ion exchange system is a significant capital investment and from the first day of operation the equipment assets will degrade. The ion exchange system will require proactive maintenance to protect the initial capital investment and ensure that it continues to perform as designed. This chapter provides an overview of key aspects of the maintenance activities associated with ion exchange systems.

With advancements in computer technology, much of the operational activities have been automated; however, there is still significant manual operation needed. Manual labor is needed to prepare the regenerant chemicals, initiate and monitor regeneration and backwashing, collect samples, and calibrate instruments.

Operational activities have costs that, if not properly managed, can escalate to levels not anticipated during the planning stages of the project. The most significant costs in ion exchange systems are associated with resin replacement, regeneration chemicals, and regenerant disposal.

The previous chapters have identified equipment and treatment processes that can be used to treat various contaminants. This chapter is intended to provide a high-level understanding of the monitoring, commissioning, and operation of those systems. The following topics are discussed:

- Instrumentation—identification and selection of appropriate monitoring equipment to monitor the treatment process.
- Commissioning—the act of confirming that system components are checked, calibrated, and functioning as designed. Starting the system and training personnel is also included in the commissioning process. The details of this process are described in a "commission plan".
- Operations—in order for the system to operate as designed, there is a need for periodic maintenance, calibration confirmation, and documentation of that maintenance. Standard operating procedures (SOPs) are developed during the commissioning process and used to train personnel on the proper operational requirements.

14.2 Instrumentation

The previous chapters described removal of the following contaminants:

- Hardness
- Total organic carbon (TOC)
- Radionuclides
- Arsenic

- Nitrate
- Perchlorate
- Chromium
- per- and polyfluoroalkyl substances (PFAS)

The removal of these contaminants using ion exchange is performed on a batch basis and when the resin has reached its capacity, it is either regenerated or replaced. As mentioned above, regenerating the resin or replacement of the resin (for single-use resins) is one of the most expensive operating parameters. Determining the optimal time to regenerate or replace the resin is an important operational aspect. As described in the individual contaminant chapters, the resin needs to be regenerated or replaced when breakthrough of the contaminant occurs and is found in the treated water at unacceptable concentrations. Prior to about the 2000s, on-site water quality instruments for these contaminants were either not available and/or were too expensive to reliably measure the feed and effluent water quality. Only laboratories had the ability to accurately measure the contaminants of concern. Operation is typically based on limited feedwater samples analyzed by a lab and measurement of the total volume treated by the ion exchange system. The capacity of the resin is known, and with the feedwater contaminant concentration, an approximate throughput volume is computed based on the resin capacity. Regeneration or replacement is performed prior to reaching the calculated throughput volume. Surface water or other sources with variable water quality will increase the need for sampling to accurately monitor throughput and maximize life of the resin. Effluent samples are also periodically collected and analyzed by a laboratory to confirm when to regenerate or replace the resin.

While resin regeneration and replacement are key operational components that benefit from instrumentation, so is the pretreatment process. As discussed in the individual contaminant chapters, ion exchange systems may require pretreatment to mitigate fouling and premature exhaustion of the resin exchange sites. On-site instrumentation in the form of continuous, bench type, and handheld instruments can help ensure that feedwater to the ion exchange system is of sufficient quality to protect the resin. While the type of instrument would depend on the specific pretreatment requirements, consideration should be given to the use of the following on-line and/or bench type instruments:

- pH, as this will assist in understanding changes to the raw water and will be impacted by ion exchange systems
- Conductivity, as this will provide an indication of the total ions in the water
- Turbidity of the raw water and downstream of any pretreatment process

Today, there is a wide range of instrumentation available for on-site use both in the form of online analyzers and bench type or handheld instruments. Online instrumentation is integrated into a treatment plant's supervisory control and data acquisition (SCADA) system.

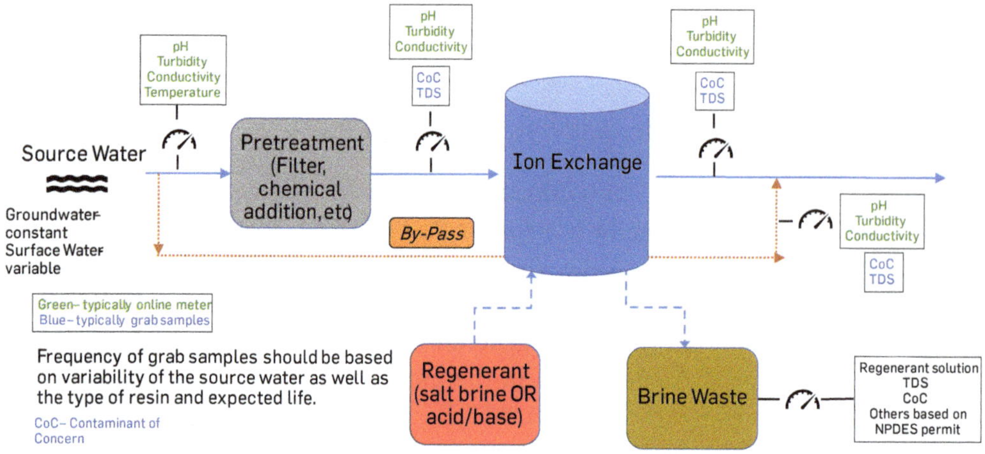

Figure 14-1 Instrumentation/sampling locations

Note: Ion exchange vessel may have sample ports that allow sampling within the bed, and consideration should be given to periodically obtaining samples from these locations.

Figure 14-1 presents a schematic of typical measurement locations of different water quality parameters and if they are typically measured by an online instrument or a grab sample to be measured with benchtop equipment or by a laboratory. Instrumentation for some contaminants of concern are undergoing development. Coordination with the instrument suppliers is beneficial to determine if a handheld or online instrument is available for measuring the contaminant of concern. The use of the instrument should consider issues such as:

- the detection level,
- accuracy,
- maintenance requirements,
- reagents,
- waste flow,
- cost, and
- history of use on similar applications.

Sampling frequency should be determined based on the regulatory requirements and how the data is used for operations.

14.3 Commissioning

The goals of the commissioning phase are to verify that:

- the design works as intended,
- it was constructed properly, and
- regulatory objectives are met.

In order to meet these goals, the facility will initially be run to waste and water quality tests performed to confirm that the system meets the water quality requirements including satisfying not only the removal of the specific ion but that the ion exchange systems satisfies microbial aspects of the safe drinking water act.

This section steps through general principles and considerations for commissioning an ion exchange system.

Pretreatment System—if applicable

Pretreatment is an important component of the commissioning phase. As a minimum, most ion exchange systems are equipped with cartridge or bag filters, and these systems should be operational prior to commissioning of the ion exchange system. Spare bags and cartridge filters should be available during commissioning, and the pressure drop across the filters should be monitored during the commissioning phase.

Additional pretreatment unit processes may be provided depending on the raw water quality, and these additional unit processes should be operational prior to commissioning of the ion exchange system.

Failure of adequate commissioning of pretreatment can catastrophically damage the IX resin, and this is especially relevant for chemicals that are fed upstream of the ion exchange system. Depending on the pretreatment process chemical feed system, there may be an adverse effect or even a reaction with the ion exchange resin.

Pretreatment is very important for single-use ion exchange systems, especially for suspended solids filtration, when the ion exchange system is operated in a downflow manner. Suspended solids can blind off resin beds and create an unacceptable pressure drop, limiting flow and performance.

With ion exchange systems that are regenerated, there is a backwash step in the regeneration sequence that will eliminate most suspended solids; therefore, pre-filtration is less important or not needed.

If the pretreatment system includes chlorine, then consideration should be given to the chlorine residual in the effluent of the pretreatment process. Ion exchange resins have a low tolerance for chlorine, and the chlorine residual of the pretreatment effluent should be measured continuously. Some pretreatment systems include a chlorine removal process, and during commissioning, the performance of the dechlorination system should be evaluated in detail.

14.3.1 Pre-startup Guidance

The pre-startup activities are guided by the design specifications and may include a punch list and a "walk-through" of the system, stepping through each process component for quality and correct installation. It is typically performed by the contractor, and the engineer and owner may be present and participate. The following is a list of key components of an ion exchange system that should be checked as part of the pre-startup procedures.

Resin vessels

Potable water is used in checking the integrity of the resin vessels. If available, use a service hydrant for checks. If potable water is not available on-site, it may need to be trucked in for the test.

- Pressure testing the vessels and piping for leaks
- Internals checking for mechanical integrity
- Instrument calibrations and factory acceptance testing of the programmable logic controller (PLC)
- Sanitization per ANSI/AWWA C653-13, *Disinfection of Water Treatment Plants*

Regeneration system

- Chemical feed pump calibrations with water
- Chemical tank leak testing with water

Waste handling

- Leak testing of waste tanks with water
- Instrument calibration

For all of these components, verification of electrical installation and checking of motors for rotation, power, etc., should also be conducted.

14.3.2 Startup

System startup should begin with a comprehensive walk through of the site with all parties. Any hazards or safety considerations should be identified. The following is a list of items that should be checked during start-up.

Resin loading and commissioning

- Packaging
  - Supplier specific: supersacks, drums, and 1 ft^3 bags
  - Slurry truck delivery
- Resin loading
  - Before loading the resin, make sure there is a water cushion at the bottom of the vessel. For initial fill, chlorinated potable water may be acceptable—coordinate with the resin vendor with respect to the compatibility of the resin with chlorinated water.
  - Loading methods include
    - gravity fill top (manual),
    - slurry for large systems,
    - eductor (water flow, create a vacuum), and
    - portable air compressor and diaphragm pump for smaller systems or in remote locations.
- Backwash if required by the manufacturer based on the type of media, as some ion exchange media require backwashing.
- Conditioning of the media is required and may entail regeneration of the system one or more times. The designer can specify that the ion exchange resin is supplied in the functional form that will be used

during normal operation to minimize the amount of work during commissioning. As an example, if the treatment process is water softening, which will be accomplished using NaCl as the regenerant solution, the designer can specify that the resin is initially supplied in the Na+ form.

Once the resin is installed in the pressure vessel and the mechanical and electrical components have been checked, the system is ready to be started for the first time. For standard fixed bed, down flow configurations the system is typically run to waste for 5–20 bed volumes, and a sample of the effluent is taken to verify the system is achieving the expected water quality results. The following water quality samples should be taken during this run to waste period:

- Total coliform (effluent)
- Specific contaminant that the ion exchange system is designed for (for example hardness, chromium, arsenic, perchlorate, radium, uranium, TOC). Influent and effluent samples should be obtained
- pH, TDS (influent and effluent)

Regeneration System—if a regenerable resin is used

- Pre-startup
  - Regenerant solution pumps—calibrate with water
  - Finalize the concentration of brine solution
  - Verify the dosage of brine that is to be used for the regenerant solution pumps
  - Install the regeneration chemicals in the brine storage tanks
- Operation
  - Monitor brine concentration, flow rates, dosage
  - Monitor and confirm throughput per design
  - Elution study of regeneration system
    - An elution study measures salt concentration and brine flow rate over time to evaluate the performance of the ion exchange regeneration system. Comparing the collected data to engineering calculations can help identify problems with the resin or the equipment. For example, fouled resin may give short runs even though the correct dose of salt has been applied.

Figure 14-2 provides an example of a commissioning trailer that contains all of the equipment required to fill vessels with resin, rinse and regenerate installed resin, and recirculate chlorine for system disinfection.

14.3.3 System Performance Test

The intent of the system performance test is to confirm all the moving parts work as intended, all components have been integrated correctly, and that the ion exchange system is operating as intended before the ion exchange system is turned over to the owner, at which time the owner will be responsible for system operation and the warranty period will begin. If applicable, performance of the

Figure 14-2 Example of a commissioning trailer
Courtesy of Water Remediation Technology LLC (WRT)

pretreatment system will also have to undergo a performance test. A performance test plan should describe roles and responsibilities between owner, engineer, contractor, and vendors. A report summarizing the test results should be prepared. Consequences of failure to achieve performance goals should be outlined in the specifications.

The system performance test will be specific to each ion exchange facility, but at a minimum, the performance test should test the following:

- Performance duration: for regenerable systems, this should describe the number of regeneration cycles. For single-use resin systems, a minimum of two weeks should be considered due to the time associated with receiving water quality results from a certified laboratory.
- Range of flow rates: consider performing tests to demonstrate that the plant can operate at the entire range of flow rates defined in the specifications.
- Water quality requirements:
  - Feedwater quality should be clearly defined. If multiple feed sources/wells are part of the project, then the water quality requirements should reflect any differences in source water quality.
  - Any wastewater sent to the sewer must meet the requirements of the receiving wastewater treatment plant.
  - Treated water sampling parameters and frequency should be defined to confirm the system is treating water to the required levels.
- Maximum head loss as defined in the specifications
- Regeneration system including chemical usage, and waste volumes including backwash and regeneration wastes

- Wastewater quality including the quality of the backwash and regeneration wastes

14.4 Operation

After the ion exchange has been started and passed all system performance test requirements, the ion exchange system will be turned over to the owner for routine operation and maintenance. Although the owner is responsible for proper maintenance, there is typically a warranty provided by the equipment supplier. The responsibilities of the owner and suppliers during the warranty period should be clearly defined as part of the procurement process.

14.4.1 Routine Operation

The operation of an ion exchange system will be based on changes in demand and changes in source water quality. For regenerable systems, regeneration is typically performed based on total throughput at which time the regeneration system is initiated and operation of the valves and pumps are automated. The plant operators will change setpoints on the control system to account for changes in demand and source water quality. Depending on the target contaminant, on-line and real-time monitoring of the treated water contaminant concentration may be available. Real-time monitoring of the treated contaminant level can provide a more efficient operation, as the system is not prematurely regenerated and can provide a more consistent treated water discharge level. Even though the ion exchange system is automated, routine operation should include the following components.

14.4.1.1 Water Quality

Perform an elution study (see Section 14.3.2) of the regeneration system on a schedule (e.g., quarterly and if a reduction of performance is noticed).

Collect and analyze samples at least weekly from the feed and effluent of the ion exchange system. The frequency should be coordinated with the regulator. Additional locations should be considered based on the extent of any pretreatment system.

14.4.1.2 Hydraulics

Manual readings on a daily basis should be taken for the flow meters and pressure gauges in order to supplement the monitoring being performed by the control system. Totalized flow should also be taken to determine how many bed volumes have been completed. This will help the operator project media replacement and regeneration requirements. These manual readings will enable the operators to identify potential issues with the performance of the ion exchange system and rectify them before they become major issues.

14.4.1.3 Alarms

The control system will generate alarms related to issues such as level, flow, pressure, pump engagement, and valve position. Standard operating procedures should be developed to respond to each alarm, and a troubleshooting guide should be provided by the equipment supplier.

14.4.1.4 Reports/Documentation

The data generated by the control system, manual readings, and laboratory analyses should be logged, trended, and analyzed. The online measurements should be stored permanently, as this data may be needed as part of any warranty claims and as part of any system optimization. The water quality data and hydraulic data should be trended/graphed over time, and the data should be reviewed with the project team on monthly basis or more frequently as needed.

14.4.2 Maintenance

An ion exchange system is an important asset to a water treatment facility and needs to be properly maintained. Maintenance requirements will be specific to each system, but the list below provides an example of what should be included:

- System maintenance requirements
  - Different components will need maintenance on different schedules: daily, weekly, annually. It is important that the instructions of the manufacturers be followed and logged in order to maintain the warranty and to help troubleshoot if issues arise.
- Water quality testing/calibration requirements
  - Different instruments will need maintenance on different schedules: daily, weekly, annually. It is important that the QA/QC procedures of each instrument be followed and logged. Failure to properly calibrate instruments can result in operational issues.
- Annual resin analysis—a resin core sample should be obtained annually and sent to the resin supplier for analysis. The supplier will comment on resin health. Resin analysis may include the following:
  - Total capacity: this analysis describes the exchange capacity of the sample and compares it to the capacity of a comparable new resin. Capacity is used to judge a resin's extent of deterioration and to determine when a resin replacement is advisable. The difference in total capacities does not always translate to operating capacity differences. At 20 percent less capacity than specification, consider replacing the resin.
  - Moisture content: this analysis describes the percentage moisture holding capacity (MHC) of the resin beads and compares them to what would be expected in a newly provided comparable resin. An increase in MHC indicates a loss in cross linking, which might be caused by oxidation of the resin beads. A decrease in MHC indicates fouling of the resin bead or deactivation.

- Percent regeneration: This test is similar to the total capacity test except it is performed on the resin as received without full conversion to a reference ionic form, which effectively shows how efficient the regeneration schedule is. A typical range for a sample from the field is 65–80 percent; any lower indicates a problem with the regeneration operations.
- Bead integrity: this analysis describes the amount of whole, cracked, and broken beads as a percentage of the total volume. Whole beads show no flaws or cracks. Cracked beads will face higher risk of bead fragmentation. Broken beads are the beads that have broken into two or more pieces, which will fill the column void spaces resulting in higher pressure drops across the resin bed. When there is an increase in broken beads of 15 percent or more, consideration should be given to replacing the resin.
- Particle size distribution: The particle size distribution describes the bead distribution on a sample of resin based on the percentage of cumulative volume. The distribution can provide information on the presence of fines, the mean diameter of the sample measured, and the uniformity of the distribution. The analysis method should be coordinated with the resin supplier.
- Metal impurities: The results are reported in ppm on wet resin and can help determine resin foulants or regeneration efficiencies. For H form cations:
  - <400 ppm: considered low metal content
  - 400–3500 ppm: considered moderate metal content
  - >3500 ppm: considered heavy metal content
- Organic fouling of anion resin: The strong base anion groups are what determine the resin's ability to remove weak anions such as silica and carbon dioxide. Significant loss of strong base % is usually traceable to organic fouling and sometimes to thermal degradation, and leads to poor resin performance, especially with respect to removal of silica.

CHAPTER 15
Case Studies

Case studies are provided for DOC, nitrate, PFAS, and radionuclides. The case studies describe projects where ion exchange has been implemented for the removal of these contaminants. The case studies include a description of the plant capacity, the raw water quality characteristics, and a description of the ion exchange system that was used.

CASE STUDY: DISSOLVED ORGANIC CARBON (DOC)

Overview

Contaminant of concern	Dissolved organic carbon (DOC)
Treatment goal	30% reduction in UVA254 (m^{-1})
Location	Adjacent to the Peak District National Park, close to Sheffield, UK
Installation year	2019
Municipality/utility	Yorkshire Water Services
Water source	Surface water
Flow rate	60 MLD (15.9 MGD)
Special considerations e.g., site constraints,	Located adjacent to the Peak District National Park, spare site space very limited
Resin type selected	SBA—MIEX
Reason for selection	Pretreatment for DOC reduction met UVA requirements and improved solids loading on filters by 33%
Treatability study	Yes
Piloting	No—Client had another WTP with the same process located nearby and on similar source

Raw Water Quality

pH	5.0 – 7.2
Alkalinity	<20 mg/L as $CaCO_3$
TSS (turbidity)	1.4–19.4 NTU
TDS	Not routinely measured
Contaminant of concern	DOC
Competing ions	None
Other parameters	Sulfate—not routinely measured
Iron	0.652–3.108 mg/L
Manganese	0.033–0.818 mg/L
LSI	Negative

Design Parameters

Flow rate	40 MLD avg; 60 MLD max (10.6 MGD avg; 15.8 MGD max)
Bypass	Variable—as required
Pretreatment	Band screen (6 mm)
Number of treatment vessels in service	Dual contacting chambers each with 9.7 m^2 (104 ft^2) surface area
Number of vessels on standby	None required
Linear velocity	13.3 m/h at max. flow (5.4 gpm/ft^2)
Specific flow rate	29.8 BV/h at max. flow (3.7 gpm/ft^3)
Empty bed contact time	2.6 min at 47 MLD (12.4 MGD) and 2.0 min at 60 MLD (15.9 MGD)
Treatment rate (bed volumes, BV)	650 BV max; 800 BV avg
Number of regeneration vessels	3
Number of regeneration vessels on standby	1

Enhanced Brief

The Langsett Water Treatment Works is located adjacent to the Peak District National Park and has a design capacity of 60 MLD (15.9 MGD). The site available spare space is extremely limited; therefore, all changes implemented on site to increase flow or to improve water quality must be compact and efficient. The source water for the facility is an upland reservoir with DOC ranging from 4 to 23 mg/L. Due to a deterioration in raw water quality over the years since the works was originally built, it was not possible to run at maximum throughput during periods of poor raw water quality with the original site treatment processes. This prevented the site from making full use of its abstraction licence and resulted in impaired operating efficiency and a reduction in resilience across the regional water network.

Yorkshire Water Services (YWS) had prior experience with fluidized anion exchange pretreatment at several of their other Treatment Works, and bench studies indicated that this process may afford enough of a reduction in UVA to allow the utility to meet DBP regulations using sodium hypochlorite for disinfection while lowering the dose of ferric sulfate. YWS used its past experience operating fluidized bed ion exchange as pretreatment to coagulation, together with coagulation bench testing at the Langsett site, to predict the new solids loading rate required to achieve both water quality targets and requisite filter loading rates.

The following water quality improvements were achieved by construction of a 60 MLD (15.9 MGD) Fluidized Anion Exchange MIEX system at the Langsett WTW:

- 30% reduction of UVA
- Reduction in ferric sulfate coagulant of 31–52%
- Reduction in solids loading rate commensurate with reduction in ferric sulfate
- Increased water production from the WTW during high DOC seasons of up to 33%
- Decrease in chlorine consumption of 20–30% approx. for same residual
- TTHM (THM4) in final treated water of 6–26 µg/L for 2020
- HAA9 in final treated water of 7–22 µg/L for 2020

The following graphs show the reduction in THM4 and HAA9 in 2020 over previous years after the treatment system was installed in 2019.

The following were the results of this process optimization:

- Residual brine is trucked off site into the sewage treatment facility.
- Resin consumption / use: on average flow of 40 MLD (10.7 MGD), 0.3–0.5 L/ML; at maximum flows of 60 MLD (15.0 MGD), 0.7–0.8 L/ML.

This is the fourth Fluidized Bed Anion Exchange Pretreatment Systems (all MIEX systems) installed and operational at Yorkshire Water Services. The Langsett WTW was the first that utilized the High-Rate process model in order to fit the tight site requirements.

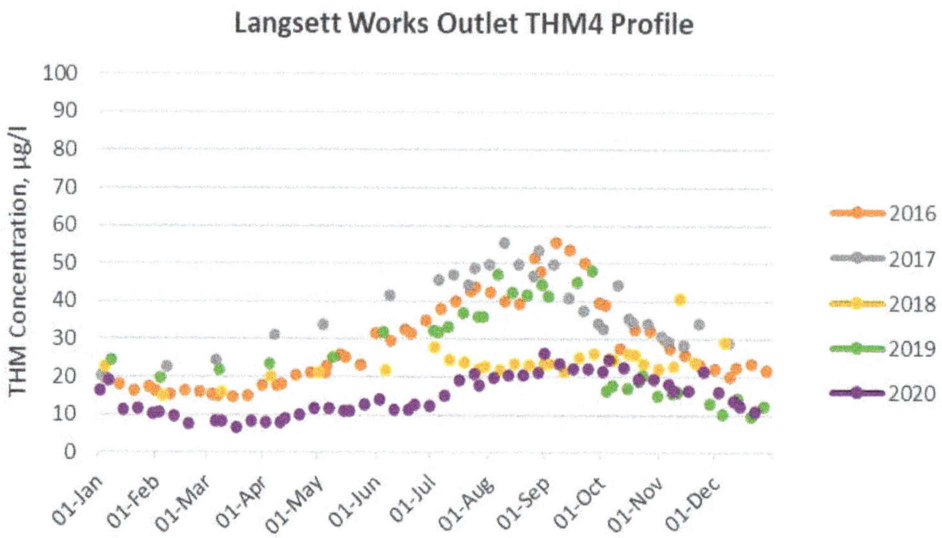
Langsett Works Outlet THM4 Profile
THM Concentration, µg/l
100
90
80
70
60
50
40
30
20
10
0
01-Jan
01-Feb
01-Mar
01-Apr
01-May
01-Jun
01-Jul
01-Aug
01-Sep
01-Oct
01-Nov
01-Dec
2016
2017
2018
2019
2020

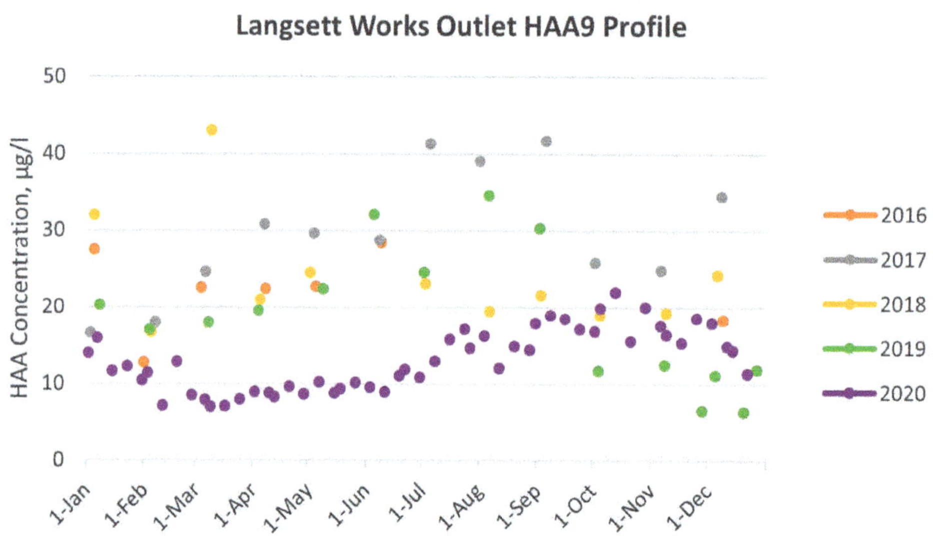
Langsett Works Outlet HAA9 Profile
HAA Concentration, µg/l
50
40
30
20
10
0
1-Jan
1-Feb
1-Mar
1-Apr
1-May
1-Jun
1-Jul
1-Aug
1-Sep
1-Oct
1-Nov
1-Dec
2016
2017
2018
2019
2020

Photo 1 Yorkshire Water Services Langsett WTW MIEX Building exterior showing salt saturators and coagulation/floc sed basins. Photo courtesy of Sasa Golubovic.

Photo 2 Yorkshire Water Services Langsett WTW MIEX Regeneration Vessels. Photo Courtesy of Sasa Golubovic.

Photo 3 Yorkshire Water Services Langsett WTW High Rate Contactor Basins with superstructure to support mixer motors (in blue) and access points. Photo Courtesy of Sasa Golubovic.

Client Contact Information:

Yorkshire Water
Western House, Western Way,
Bradford BD6 2SZ, United Kingdom
0345 1 242424
www.yorkshirewater.com

CASE STUDY: NITRATE

Overview

Contaminant of Concern	Nitrate
Treatment goal	Blended treated water target of 23.8 mgd (90 MLD) at <42 mg/L as NO_3
	Approximately one-third of the flow 8.45 mgd (32 MLD) to be treated in the ion exchange plant, which achieves <5 mg/L as NO_3
Location	Keldgate WTW, Hull, United Kingdom
Installation year	2010
Municipality/utility	Yorkshire Water
Water source	Groundwater
Flow rate	23.8 mgd (90 MLD) treatment plant, 8.45 mgd (32 MLD) capacity of ion exchange system
Special considerations, e.g., site constraints	Waste attenuated in a buffer tank before discharge to local sewer system
Resin type selected	Nitrate selective SBA
Reason for selection	Nitrate selective resin and UK Drinking Water Inspectorate approved
Treatability study	No
Piloting	No

Raw Water Quality (Average)

pH	7.8
Alkalinity	247 mg/L as $CaCO_3$
Turbidity	10 NTU
TDS	481 mg/L
Contaminant of concern	Nitrate—design for anticipated future concentration of up to 63 mg/L as NO_3
Competing ions	Sulfate, bicarbonate

Design Parameters

Flow rate	23.8 mgd (90 MLD) treatment plant, 8.45 mgd (32 MLD) capacity of ion exchange system
Bypass	Yes, up to 8.45 mgd of 23.8 MGD (32 MLD of 90 MLD) treated by ion exchange system
Pretreatment	Boll filters for particulate removal (50 microns). The water is treated by nano filtration membranes downstream of the ion exchange plant.
Number of treatment trains and capacity per train	Three ion exchange treatment trains operating in parallel. Two in operation with one in stand-by. Each train has capacity to treat 50% of the treatment flow. Each treatment train has a capacity of 4.2 mgd (16 MLD)
Number of vessels in service	Three trains each comprising 24 vessels, of which 18 are in service at any given time with three vessels in the regeneration zone, two in the regeneration rinse zone, and one in the displacement zone.
	In each train, the patented multi-port distributor valve connects to (and from) the 24 resin vessels. A rotating process disc within the valve moves (indexes) at regular intervals to change the 'process position' of each vessel (the physical position remains unchanged).
Number of vessels on stand-by	One train comprising 24 vessels
Vessel volume	4.67 ft^3 (1.18m^3)
Linear velocity	8.6 gpm/ft^2 21(m^3/h)/m^2
Specific flow rate	3.9 gpm/ft^3 31.1 m^3/h per m^3 resin volume based on a flow of 36.7 m^3/h per in service vessel
Empty bed contact time	1.93 min
BV/h	31.1 BV/h

Enhanced Brief

The Keldgate Water Treatment Works (WTW), which is owned by Yorkshire Water, draws groundwater from natural borehole supplies. Tests indicated that the nitrate level in the water supply had been increasing steadily for some time. While prevailing concentrations were acceptable, the feedwater nitrate concentration was predicted to reach 63 mg/L as NO_3 by 2025, which exceeds the regulatory limit in the United Kingdom for nitrate 50 mg/L as NO_3. It was clear Yorkshire Water had to act to mitigate the possibility of breaching the regulatory limit.

During feasibility studies, it was determined that discontinued use of high-nitrate boreholes would not leave the works with sufficient yield of raw water. Similarly, importing water for blending from low-nitrate sites was discounted as further treatment would still have been required. A preferred option was identified to construct a new nitrate removal plant at Keldgate WTW.

Initially several technologies were considered, including:

- reverse osmosis,
- electrodialysis reversal, and
- ion exchange.

Ion exchange technologies were shown to offer the most cost-effective solutions in terms of both capital and operational costs. They also offered the benefit of a much-reduced waste stream compared with other technologies. This was a significant factor in the decision-making process given limited capacity in the local sewer network.

Ion exchange resin is vulnerable to chlorine attack. Mitigation options included:

- location of the ion exchange plant upstream of the pre-membrane chlorination dosing point, and
- location of the ion exchange plant downstream of the membranes with de-chlorination upstream and re-chlorination downstream

With a fixed bed ion exchange system, there were concerns that periodic high concentrations of fine chalk particles would lead to fouling of the resin. This might suggest the most appropriate location of the ion exchange plant was downstream of the membrane plant. However, the chosen ion exchange technology includes an innovative fluidized bed process, which was considered to significantly reduce the risk of fouling of the resin and allowed the ion exchange treatment to be located upstream of the existing membrane plant. This negated the need for de-chlorination and re-chlorination, resulting in significant cost savings.

The Keldgate facility has a capacity of 90 MLD, of which 32 MLD can pass through the ion exchange nitrate removal system. Treated water is blended back into the main flow and will allow for nitrate removal to 42 mg/L as NO_3, well below the regulatory level of 50 mg/L as NO_3.

This facility was constructed by Mott MacDonald Bentley (MMB) and uses a novel treatment process developed by ACWA Services called NITREAT. The

NITREAT continuous counter current ion exchange system is an improvement on conventional "batch" treatment systems. It enables exhausted resin to be continuously regenerated, rinsed, and returned to work, with all phases of the cycle occurring simultaneously.

This process uses a larger number of smaller vessels in parallel. This approach provides several benefits. The small pressure vessels allow for better utilization of the resin because the water is more effectively distributed through the resin bed and minimizes channeling. The large number of vessels allows for continuous operations versus a batch process of taking a column offline for regeneration and putting in back in service. With the NITREAT process, the vessels operate in four different zones: adsorption, displacement, regeneration, and regeneration rinse. At the Keldgate WTW, there are a total of three trains, each with 24 vessels per train. In each train, 18 of the 24 of the vessels operate in the adsorption zone, in-service treating water. The remaining six vessels are in the stages of regeneration. One vessel is continuously taken out of service while a regenerated vessel moves back into the adsorption zone. The process as a whole operates as a continuous process, and there is no disruption to the treatment to regenerate resin.

Waste flow from the NITREAT process is relatively small thanks to the recovery and reuse of the process streams with waste volumes of less than 0.5 percent of the treatment plant's output. The solution also included the capacity to store all waste flows for a minimum of 12 h in the event of the receiving public sewer surcharging (an automated system inhibits discharge from the site when levels in the downstream sewer exceed agreed levels).

Photograph

3D Drawing of the Treatment Plant, Courtesy of ACWA Services Ltd.

The plant has been in successful operation since 2010, achieving all of its performance criteria with no major issues.

Client Contact Information
Yorkshire Water
Western House, Western Way,
Bradford BD6 2SZ, United Kingdom

CASE STUDY: PFAS

Overview

Contaminant of concern	PFAS (including PFOA and PFOS)
Treatment goal	Non-detect levels (~2.5 ppt)
Location	Horsham, Pa.
Installation year	2017
Municipality/utility	Horsham Water & Sewer Authority
Water source	Groundwater
Flow rate	50 gpm
Special considerations, e.g., site constraints,	Temporary unit, vessels needed to fit into small existing well house
Resin type selected	Purolite–Purofine PFA694E PFAS-Selective Resin
Treatability study	Yes
Piloting	Yes

Raw Water Quality

pH	6.5 units
Alkalinity	85 mg/L as $CaCO_3$
TSS	0.5 mg/L
TDS	Conductivity 460 uS/cm
Contaminant of concern	PFAS (including PFOA and PFOS)
Other parameters	Chloride (60 mg/L), sulfate (20 mg/L), nitrate (2.8 mg/L as N)

Design Parameters

Flow rate	50 gpm
Bypass	0%
Pretreatment	Pre-sedimentation cartridge filter
Number of vessels in service	Originally 2, now 3 in series
Number of vessels on standby	None
Vessel volume	20 ft^3 resin
Linear velocity	10.2 gpm/ft^2
Specific flow rate	2.5 gpm/ft^3
Empty bed contact time	1.8 min
BV/h	20 BV/h

Enhanced Brief

The Horsham Water and Sewer Authority (HWSA) system has 14 wells, all contaminated with PFAS because of the historical use of aqueous film forming foam (AFFF) at the Willow Grove Joint Naval Air Station located in Horsham, Pa. While five of the wells have levels above the existing 70 ppt PFOA/PFOS Health Advisory Level and PFAS removal treatment for these wells are being funded by the Navy, Horsham Township directed the Authority to install treatment on all sources to help mitigate any potential long-term health impacts from the decades of past exposure to the community. While the need for quick installation of treatment necessitated initial permit requests to specify granulated activated carbon (GAC) filters, with HWSA anticipating years of treatment at a large number of sources, the Authority felt it necessary to investigate alternate treatment methods as well. As such, one well (Well #10) was selected to pilot anion exchange resin.

From the beginning, the permit approval process for the ion exchange has been more challenging than for GAC. The majority of available literature on PFAS removal is based on GAC applications, so there is more comfort with that technology in some circles. Although PFAS is not regulated by USEPA, the state regulators in Pennsylvania deemed ion exchange removal of PFAS as an "innovative technology" and placed more hurdles in the pilot permits than for any of the GAC permits. HWSA took several measures to address regulator concerns, including the following:

- Recruited a home in close vicinity to Well #10 with confirmed lead solder in its kitchen. Technically, this home and all the others in the neighborhood of Well #10 are Tier 2 homes under the current Lead and Copper Rule (LCR) and cannot be used as an LCR compliance point as they were built before 1982. Performed multiple LCR protocol stagnation sampling at this home to show that lead was not an issue.

- Performed daily monitoring of chlorine, alkalinity, pH, and temperature, as well as weekly monitoring of chloride and sulfate, for Well #10 water quality parameters for more than one year to show their stability following ion exchange treatment.
- Extensive testing when a third vessel was added (as will be described here).
- The resin manufacturer, Purolite, showed the regulators its water chemistry model.

GAC pilots required one month of operation essentially to verify that the application would adequately remove the PFAS. The ion exchange pilot at Well #10, however, has been in operation for more than three years, with much of the concern related to potential simultaneous compliance issues described here.

The state regulators have been concerned about the potential for the ion exchange resin to impact the corrosivity of the water due to potential changes in the chloride-to-sulfate mass ratio (CSMR). Despite assurances of the resin manufacturer that any changes would only occur in the initial start-up of a new vessel but quickly stabilize back to background levels, substantial monitoring has been required to verify those claims. Graphs of the testing performed following the installation of a third vessel to our pilot is provided at the end of the case study. This testing showed that although there are initial impacts to the alkalinity, chloride, sulfate, and nitrate levels, they all return to background conditions quickly and remain constant such that the impacts can be negated through an initial rinse to waste procedure for new resin.

The application uses a single-use resin; therefore, no regeneration is performed. Additionally, unlike with GAC, there is no backwash needed. As such, there is no waste stream involved in the process other than the initial rinse to waste anticipated to alleviate the simultaneous compliance concerns. To date, only one vessel (20 ft^3) of resin material has been disposed of. The resin was incinerated, and the ash was disposed of at the double-lined landfill.

Based on the piloting, ion exchange will result in lower capital cost (although the resin is more expensive per pound than GAC, less material is needed for the same flow rate and thus also requires a smaller footprint) and operating costs (material change-outs are expected at greater than three year intervals compared to the approximately one year intervals for GAC). There is also no PFAS-laden waste stream issue to deal with as there is with GAC. Although there are impacts to the anion concentrations of the water during start-up, they can easily be negated through a rinse-to-waste procedure.

The system has been performing as expected since installation.

Photo 1 Well #10 with Two IX Vessels

Photo 1 Well #10 with Three IX Vessels

Water Quality Testing Graphs at Start-Up for Vessel #3

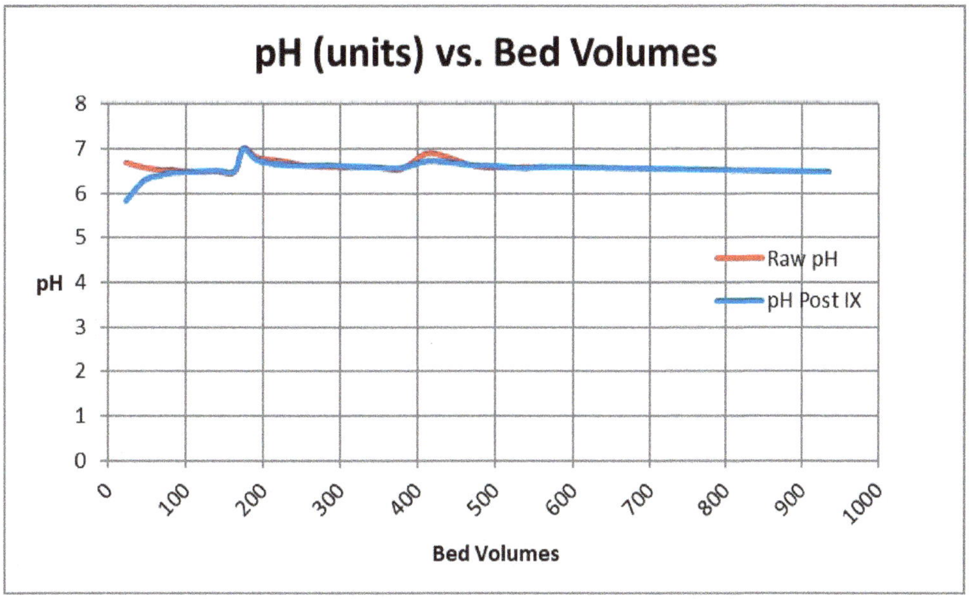

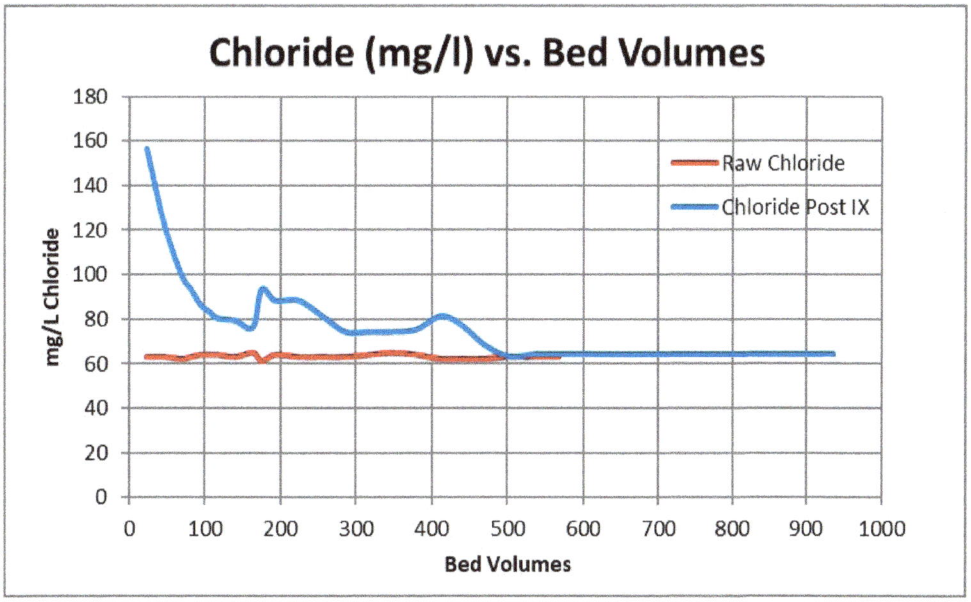

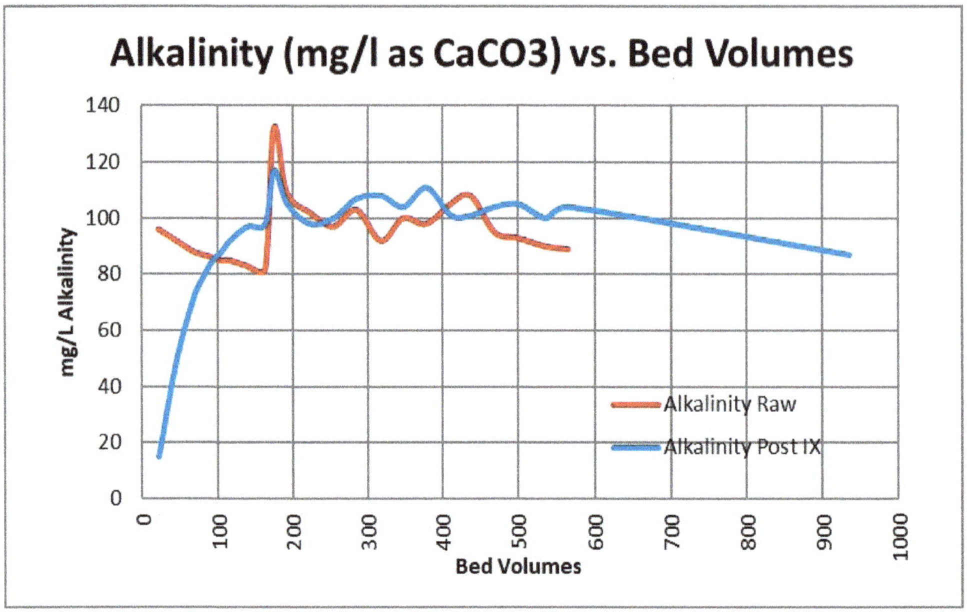
Alkalinity (mg/l as CaCO3) vs. Bed Volumes
mg/L Alkalinity
140
120
100
80
60
40
20
0
0
100
200
300
400
500
600
700
800
900
1000
Bed Volumes
Alkalinity Raw
Alkalinity Post IX

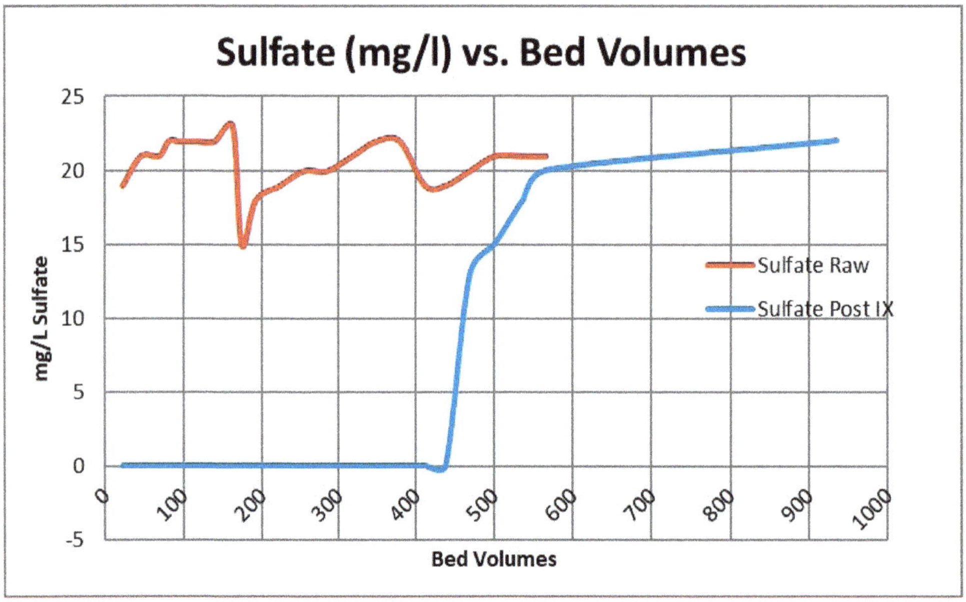
Sulfate (mg/l) vs. Bed Volumes
mg/L Sulfate
25
20
15
10
5
0
-5
0
100
200
300
400
500
600
700
800
900
1000
Bed Volumes
Sulfate Raw
Sulfate Post IX

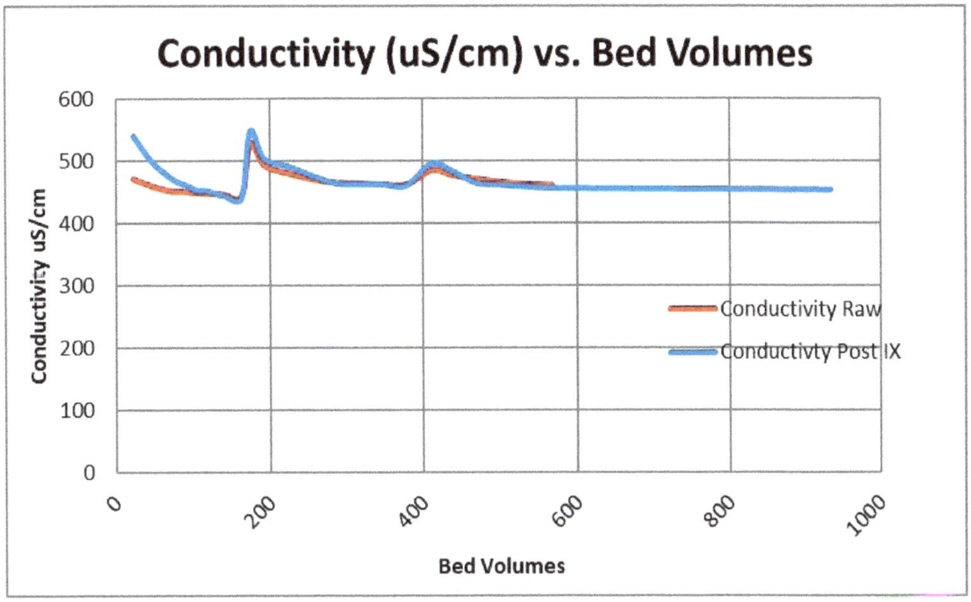

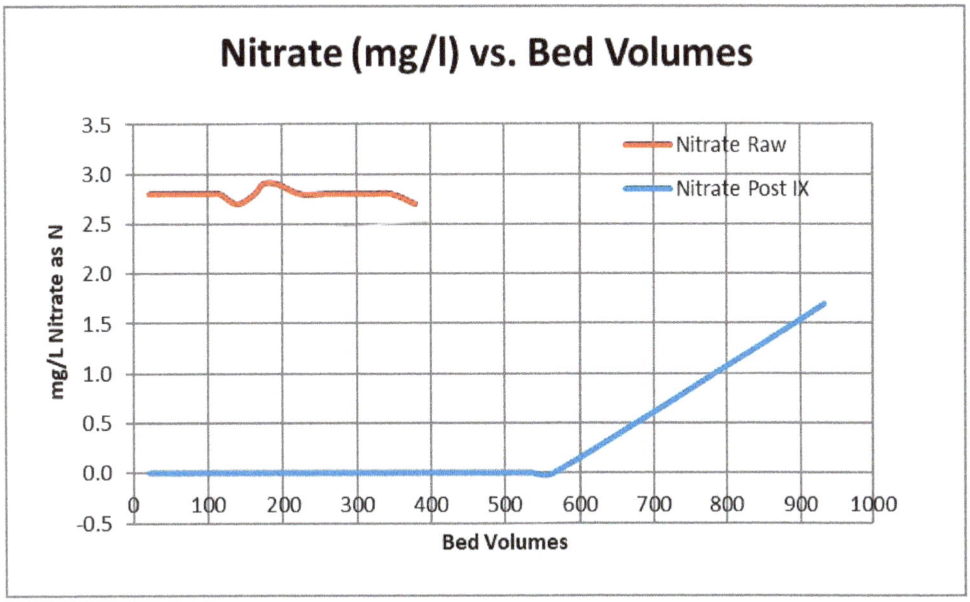

Client Contact Information

Michael J. Pickel, PE
Director of Compliance & Regulatory Affairs
Horsham Water & Sewer Authority
617 Horsham Road
Horsham, PA 19044
https://www.horshamwater-sewer.com/

CASE STUDY: RADIONUCLIDE

Overview

Contaminant of concern	Gross alpha and combined radium 226/228
Treatment goal (50% of MCL)	
Gross alpha	<7.5 pCi/L
Combined radium	<2.5 pCi/L
Installation year	2017
Municipality/utility	New Jersey American Water
Water source	Groundwater from six wells
Flow rate	800 gpm, with a maximum of 1,600 gpm
Special considerations, e.g., site constraints	Disposal of resin loaded with radionuclides
Resin type selected	Processed zeolite in cation (Na+) form
Reasons for selection	No on-site residual disposal. Licensed disposal of low level radioactive residuals.
Treatability study	No, none required
Piloting	Yes

Raw Water Quality

pH	7.4
Alkalinity	60 mg/L
TSS	<1 mg/L
TDS	86 mg/L
Influent design gross alpha	18.3 pCi/L
Influent design raw water combined radium	8.5 pCi/L
Competing ions	Barium is a competing cation, but only 0.017 mg/L in this application, so not of concern
Other parameters	Iron: Non-detect to 0.1 mg/L Magnesium: 1.5 mg/L Manganese: 0.011 mg/L

Design Parameters

Flow rate	800 gpm/train × 2 trains, for a system total of 1,600 gpm
Bypass	100% treatment, no bypass
Pretreatment	Suspended Solids Prefiltration, but no other pretreatment, or chemical feed required or recommended
Number of vessels in service	2 for each train, for a system total of 4
Number of vessels on standby	None in standby
Vessel volume	615 ft^3/vessel for a system total of 2,460 ft^3
Linear velocity	8.5 gpm/ft^2
Specific flow rate	800 gpm max per train
Empty bed contact time	11.5 min
BV/h	5.2

Enhanced Brief

To reduce gross alpha and combined radium 226/228 to below the respective MCL's of 15 pCi/L and 5 pCi/L, New Jersey American Water (NJAW) installed four (4) ion exchange radium removal vessels (two parallel trains of two vessels in series).

The project was initiated to accommodate the area population growth and the corresponding increase in water demand. A temporary system was installed on Well #13 and used for 18 months to assist with radium and gross alpha removal at the Oak Street Station until the full scale, permanent system was constructed and put in service. By operating the temporary system, it also provided technology confirmation and helped with the final permitting process.

Because the area continues to have a substantial population growth, the permanent system was designed and built with two parallel treatment trains to accommodate both the current and future water demands. The two parallel trains also provide redundancy to allow for media exchange or other maintenance without disruption to service. While one system is off-line for maintenance, the other can treat the full flow. Current operation alternates between treatment trains on alternating pump starts. Alternating treatment trains ensures that water does not sit stagnant in the resin beds for too long and cause adverse water quality conditions. As future water demand increases, the overall system can then utilize both trains on a full-time basis to handle the increase.

The corrosion inhibitor system was upgraded to accommodate the design flows.

A particular consideration with radium removal is the disposal of the media, or treatment residuals that are regulated as a low-level radioactive waste. In this case, the ion exchange equipment supplier, Water Remediation Technology LLC (WRT), provides spent media removal and licensed disposal of the spent media including all of the necessary permits, radioactive materials licensing, and spent media disposition documentation. WRT also guarantees MCL compliance, system performance, and takes ownership to the radium loaded on the media, thus mitigating risk to the utility.

From the day of commissioning, the system has operated as expected and to date has treated water discharge values to at or near non-detect for combined radium and <3 pCi/L for gross alpha.

Photo 1

Photo 2

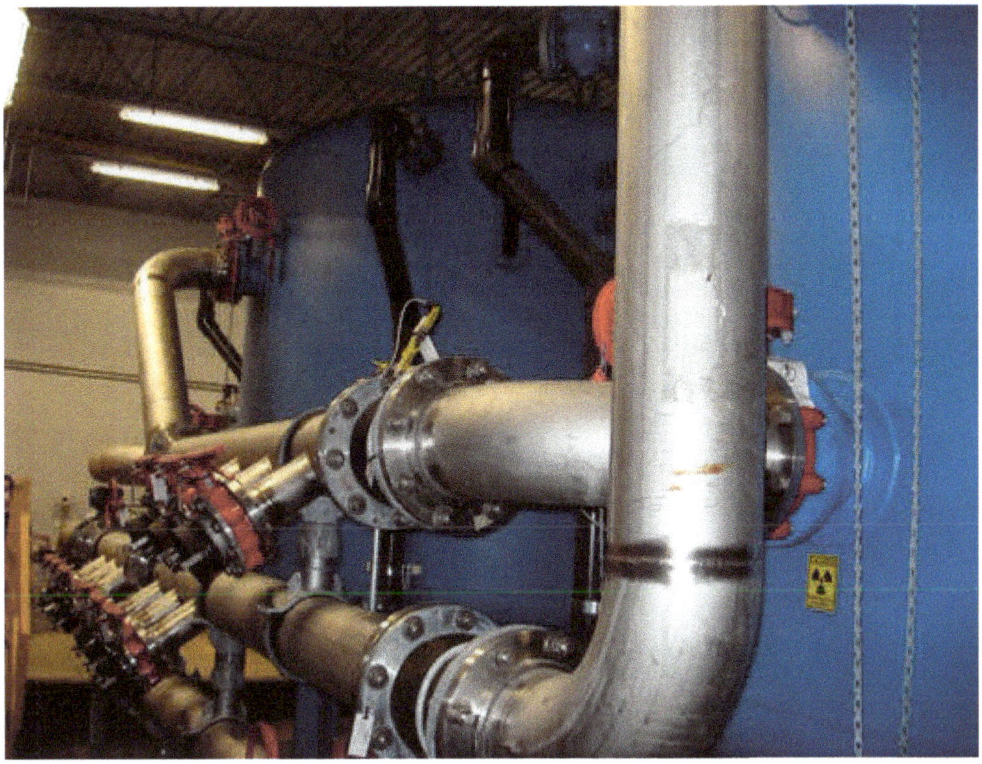

Photo 3

Client Contact Information

New Jersey American Water
https://www.amwater.com/njaw/

GLOSSARY

alkalinity The capacity to neutralize acid, mainly comprising the sum of the bicarbonates, carbonates, and hydroxides of calcium, magnesium, sodium, and potassium. Provides a buffering effect on pH.

absorption Drawing in a substance as a whole (as a sponge absorbs water).

acid A compound that yields hydrogen ions (H^+) when it dissolves in water; a strong acid is one that ionizes completely in water and yields a high percentage of hydrogen ions.

adsorption Adhesion of the molecules of a gas, liquid, or dissolved substance to a surface. In water treatment, for example, granular activated carbon (GAC) adsorbs organics to its surface.

ammonia A light, colorless gas that is soluble in water, also known as NH^3, with an irritating and pungent odor. It normally has a chemical demand for chlorine of almost eight times its own mass (i.e., 8:1). Excess ammonia in surface water discharges can cause excessive microbial growth.

anion A negatively charged ion or molecule that forms when an atom acquires one or more extra electrons.

anion exchange A processes in which anions are removed from a liquid phase by contacting a synthetic, porous medium or resin that is coated with other anions. The anions on the medium are exchanged for the anions in the solution. When the medium is depleted of the exchanging anions, it is backwashed with a concentrated solution to restore the bed with the exchange anions and to flush the removed anions. In regeneration, a concentrated solution of the virgin-form anion (like chloride) is used to reverse the equilibrium to restore the media bed with the less selective anion (put the resin back in chloride form) and flush the contaminant anions out of the bed for subsequent disposal.

anion exchange resin A synthetic material possessing the reversible ion-exchange ability for anions. Anion exchange resins are characterized as either a strong base anion (SBA) resin or a weak base anion (WBA). SBA is the most commonly used in potable water treatment and normally operated in the chloride form. SBA resin can be either regenerable or single-use depending on the contaminant. WBA is less commonly used for potable water treatment for anions.

attrition The act or process of losing some portion of the ion exchange resin due to physical or chemical degradation.

bed The treatment media inside a vessel used for ion exchange.

bed expansion The effect produced during backwashing when the ion exchange medium is fluidized and rises in the tank or column. It is usually expressed as a percentage increase of bed depth, such as 25 percent, 50 percent, or 75 percent.

bed life The time it takes for the media to lose its ion exchange capacity. When this loss occurs, the bed must be replaced or regenerated. The endpoint must be defined; for example, influent concentration is equal to effluent concentration when discussing bed life.

bed volume The volume occupied by a treatment media bed often expressed in cubic feet or cubic meters. Bed volumes is a unitless measure that indicates the capacity of a system to remove contaminant. Bed volumes are often used to measure throughput of a treatment media and are useful in scaling a treatment system. The product of the number of bed volumes and the empty bed contact time is the operation time.

bench testing Small-scale testing of chemical reactions or treatment processes on a bench, counter, or table.

breakthrough The time in the cycle of a treatment bed when an increase, sometimes defined as an unacceptable increase, in the effluent concentration for the contaminant being controlled.

brine A concentrated salt solution, generally containing sodium, chloride, and other ions typically having a concentration of 3 weight percent or more. Brine is used as a regenerant in ion exchange.

calcium carbonate ($CaCO_3$) A common substance found in rocks as the mineral calcite, most prominently in the form of limestone. It is a sparingly soluble salt, the solubility of which decreases with increasing temperature, pH, bicarbonate alkalinity and total dissolved solids (TDS). It has the potential to deposit on surfaces if it is in concentrations to super saturation.

calcium carbonate equivalence An expression of the concentration of the specified constituents in water in terms of the equivalent value to calcium carbonate ($CaCO_3$). For example, the hardness in water that is caused by calcium, magnesium, and other ions is usually described in terms of the calcium carbonate equivalent. The hardness in water that is caused by calcium, magnesium, and other ions is usually described in terms of the calcium carbonate equivalent. For example, the concentration of calcium ions (i.e., $[Ca^{2+}]$) can be multiplied by 100/40 (the ratio of $CaCO_3$'s molecular weight to Ca^{2+}'s atomic weight) to give $[Ca^{2+}]$ as equivalent $CaCO_3$.

calcium hardness The portion of total hardness caused by calcium compounds such as calcium carbonate ($CaCO_3$) and calcium sulfate ($CaSO_4$).

carcinogen A substance that causes cancer.

cation A positively charged ion, for example, calcium, magnesium, or radium.

cation exchange A process in which cation contaminants such as calcium, magnesium, or radium is removed from a liquid phase by contacting a synthetic porous medium or resin that is coated with a different cation species. The cations on the medium are exchanged for the cation contaminants. When the medium is depleted of the exchanging cations, it is regenerated with a concentrated solution to restore the bed with the exchanging cations and flush the contaminant cations for subsequent disposal (e.g., ion exchange softening where calcium and magnesium are exchanged for sodium on the cation exchange resin). In regeneration, a concentrated solution of the virgin-form cation (like sodium) is used to reverse the equilibrium to restore the media bed with the less selective cations (put the resin back in sodium form) and flush the contaminant cations (like calcium and magnesium) out of the bed for subsequent disposal (e.g., ion exchange softening where calcium and magnesium are exchanged for sodium on the cation exchange resin).

cation exchange resin A synthetic material possessing the reversible ion-exchange ability for cations. Sulfonated copolymer of polystyrene and divinylbenzene (DVB) resin is the most commonly used cation resin. It is characterized as a strong acid cation (SAC) resin and is typically used to remove both carbonate (temporary) and noncarbonate (permanent) hardness. Weak acid cation (WAC) resins contain carboxylate functional groups in place of the sulfonic acid groups. WAC will only remove temporary carbonate hardness.

caustic soda A common name for sodium hydroxide, NaOH, available in a range of solution strengths from 10 to 50 percent concentrations. It is used for pH adjustment and to regenerate strong base anion resins.

co-flow regeneration Where the service flow and the regeneration flow both flow from top to the bottom of the column. Also known as co-current regeneration.

chromatographic peaking The effect where an ion with less selectivity is concentrated on the resin. Subsequently, another ion with higher selectivity can cause the lower selectivity ion to exit the column at a higher concentration than its influent concentration. Also known as sloughing or peaking.

clean-bed head loss (*See also head and head loss*) The pressure drop across a resin bed when it is clean. If the decision to backwash is based on head loss, then equipment is triggered by the measured differential pressure across the resin bed when a set point is reached.

counterflow regeneration Where the fluids service flow (influent stream) and the regeneration stream flow in opposite direction. In this case, the regenerant does not have to push the contaminating ions through the whole resin bed. The layers that are less exhausted will be regenerated first and will be the cleanest when the next loading run (exhaustion) starts, thereby producing less contaminant leakage. Also known as counter current regeneration.

conditioning Usually, a reference to the act of applying a chemical in solution to a new, unused product, such as manganese greensand, to transform the material into a condition in which it can do its job. Before the $MnO_2(s)$ coating on manganese greensand can adsorb optimum amounts of manganese in solution, it must first be conditioned or regenerated with a strong oxidant, typically potassium permanganate ($KMnO_4$) or chlorine (Cl_2).

conductivity A measure of the ability of a solution to conduct electrical current. Conductivity can be used to estimate the total dissolved solids concentration.

contaminant Any physical, chemical, biological, or radiological substance or matter in water that is of concern undesirable for health or aesthetic reasons and requires removal.

copolymer (backbone) long-chain organic, which is ion exchange resin precursors and forms the structural component of resin. Copolymer chains are crosslinked for strength. The functional group attaches to the crosslinked copolymer backbone and determines the type of ion exchange resin. The chemical composition of most resins' backbone is either a copolymer of polystyrene and di-vinyl benzene (DVB) crosslinker (SAC, SBA, WBA) or polyacrylic DVB copolymer.

disinfection The process of cleaning that removes, kills, or inactivates microorganisms, including pathogens in water or equipment. Typical methods include adding an oxidant such as chlorine, chlorine dioxide, or ozone, or by applying ultraviolet light so that no dangerous living organisms are transmitted to humans or animals.

dissolved organic carbon (matter) (DOC) The portion of the organic carbon in water that passes through a 0.45-micrometer pore-diameter filter. For most drinking water sources, the dissolved organic carbon fraction represents a very high percentage of the total organic carbon pool. It is composed of individual compounds as well as nonspecific humic material, although humic substances account for a large portion of dissolved organic matter in natural waters. Typically, the dissolved organic carbon level provides some indication of the amount of disinfection byproduct precursors in a water source. After filtration, dissolved organic carbon is determined in the same manner as total organic carbon.

divalent ion An ion with two positive or negative electrical charges such as ferrous ion (Fe^{2+}) or sulfate ($SO^{2-}{}_4$).

dosage The amount of chemical added to raw water during water treatment, usually expressed in milligrams per liter (mg/L). For example, 0.64 mg/L chlorine (Cl_2) helps to oxidize iron; other dosages might be stated as 1 mg/L fluoride, 5 mg/L sequestrant, etc.

effective size The sieve size such that 10 percent by weight of a granular media is smaller, often shown as ES.

empty bed contact time (EBCT) Essentially, the amount of time the water takes from entry into a vessel to exiting the vessel. It can be calculated by the volume of the resin divided by the flowrate. Usually expressed in minutes.

endpoint The defined effluent concentration at which a media bed is determined to be ready for replacement or regeneration. Common endpoints include where the influent concentration is equal to effluent concentration, or 80 percent of the MCL on the lead bed.

equilibrium A condition in which reversible chemical reactions are simultaneously taking place in such a manner (i.e., at an equivalent rate) that no change in the net concentration of the substances involved in the chemical processes occurs. As a result of these simultaneous reactions, the substances involved in the reactions can be shown to be in a constant ratio to each other.

equivalent weight (eq, eq wt) A weight of a substance based on the molecular weight and number of reactive protons associated with the substance. For an element, the equivalent weight is the atomic weight divided by the number of reactive protons (valence). For example, calcium has an atomic weight of 40 and has two reactive protons (charge of 2), so its equivalent weight is 40/2 = 20. For a compound, the equivalent weight is the molecular weight divided by the number of reactive protons. For example, sodium chloride (NaCl) has a molecular weight of 58.5 and one reactive proton, so the equivalent weight is 58.5/1 = 58.5.

exchange site A location on each bead of ion-exchange resin that holds mobile ions available for exchange with other ions in the solution passing through the resin bed. In cation water softening, for example, mobile sodium ions attached to the various exchange sites throughout the resin are exchanged for calcium or magnesium ions in the water being softened. Exchange sites are also called functional groups.

fines Small particles or broken media resulting from the manufacturing process, regeneration, or natural attrition. All filter media produced by crushing contains fines, which must be removed after installation and before the filter is put into service.

finished water Fully treated water ready to enter the distribution system.

fluidization Bringing about the state in which a media behaves as a fluid because a sufficient volume of backwash water flows through the filter bed. The main factor influencing the degree of fluidization is time and backwash linear flowrate (gpm/ft^2).x

fluidization Bringing about the state in which particles such as ion exchange resin beads or sand grains become freely supported by the upward velocity of backwash water.

fouling The accumulation of undesirable foreign matter in a filter or ion-exchange media bed causing clogging of pores or coating of surfaces, channeling, and inhibition of the proper operation of the bed and the treatment system. Cause of fouling include suspended solids, bacteria, iron, manganese, etc.

functional group The ion exchange site where ions in water are exchanged for counter ions. Examples of ion exchange functional groups include sulfonic acid, amine, and carboxyl.

gel resin An ion exchange resin bead created without porosity. Often contrasted to microporous resin which has pores through it. Gel resins beads are translucent. Gel resins are not "soft" like the conventional gel definition; the nomenclature comes from the initial production stages.

hardness A measure of quality of water caused by divalent metallic cations and resulting in increased consumption of soap, deposition of scale in boilers, damage in some industrial processes, and sometimes objectionable taste. The principal hardness-causing cations are calcium, magnesium, strontium, ferrous iron, and manganese ions. Hardness may be determined by a standard laboratory titration procedure or computed from the amounts of calcium and magnesium expressed as equivalent calcium carbonate ($CaCO_3$).

hard water Water containing mineral ions that can precipitate insoluble substances. The most common hardness elements are calcium and magnesium (not to be confused with manganese). When hardness is removed by softening, the scaling potential is reduced. For example, softening allows water to more easily generate soap lather. There are no USEPA primary or secondary standards for water hardness. Water hardness is classified by the US Geological Survey as follows:

Soft	0–60 mg/L as $CaCO_3$
Moderately Hard	61–120 mg/L as $CaCO_3$
Hard	121–180 mg/L as $CaCO_3$
Very Hard	>180 mg/L as $CaCO_3$

head A measure of pressure as defined by energy possessed by a unit quantity of water at its given location, usually expressed in feet or meters.

head loss A reduction in pressure over a filter, or a resistance to flow. The major loss of head in a filtration or ion exchange system is caused by the filter media itself, since friction results from restrictions in flow. In fixed bed reactors, the smaller the resin beads, the greater the surface area resulting in higher head loss. Head loss increases as these restrictions increase or the flow passages diminish in time. Head loss varies with the depth, shape, size, and cleanliness of the media.

ion An electrically charged atom or radical.

ion exchange A chemical process for reversibly transferring ions between an insoluble solid, such as a resin and a fluid, usually a water solution. The exchange in water of the ions of one element for the ions of another. For example, sodium or potassium cations attached to a zeolite (as in a water softener) exchange themselves for cations of calcium and magnesium, softening the water passing through the zeolite bed. (*See also hard water; zeolite.*)

ion exchanger A permanent insoluble material (usually a synthetic resin) containing ions that will exchange reversibly with other ions in a surrounding solution. Both cation and anion exchangers are used in water conditioning. The volume of an ion exchanger is measured in cubic feet or cubic meters of exchanger after the exchanger bed has been backwashed and drained and has settled into place. See also *ion exchange*.

ionic bond A type of chemical bond in which electrons are transferred, formed.

ionization The splitting or dissociation of molecules in a water solution into negatively and positively charged ions or radicals.

kinetics The rate at which a reaction or ion exchange process occurs.

leakage The presence in treated water of the type of ions that the ion-exchange process was intended to remove. Because of selectivity, there is usually some low level leakage after regeneration. Removal may be insufficient due to incomplete resin regeneration, excessive service rates, low temperatures, interfering total dissolved solids, or other factors. This type of leakage is also referred to as slippage.

macroporous resin An ion exchange resin made with internal porosity and a larger pore size (100–1000 Å) than other types of resin. Pores within a resin bead run in and out of the bead like "small tunnels." Macroporous beads are opaque because the pores prevent light from penetrating through the bead.

mass transfer zone (MTZ) Portion of ion exchange media column where the treatment (ion exchange) has taken place. Consists of three parts: diffusion of the ions from the bulk phase through the liquid, diffusion of ions into the micropores of the ion exchanger, and the ion exchange reaction. The MTZ progresses down the column as the media is used up.

maximum contaminant level (MCL) A value defined under the Safe Drinking Water Act as the maximum permissible level (concentration) of a contaminant in water delivered to any user of a public water system. MCLs are legally enforced standards in the United States.

molar The number of moles per volume of solution indicated by M.

mole (mol) The quantity of a substance that contains as many molecules as there are atoms in 12 g of carbon-12.

molecular weight The sum of the atomic weights of all the atoms in a molecule.

nonionic resin Specialty solid bead–shaped, non-functionalized polymers used for adsorption of organics.

organic chemistry The chemistry of carbon compounds. Chemists have identified literally millions of carbon compounds. The terms *organic complexing* and *organic binding* refer to the process through which such elements as iron and manganese become part of a carbon compound.

organic compound A compound in which carbon is the chief constituent. organisms. A carbon-containing compound that is derived from living organisms = (e.g., a hydrocarbon, alcohol, ether, aldehyde, ketone, carboxylic acid, or carbohydrate).

oxidant A compound that causes oxidation. Examples are oxidants used for disinfection are chlorine, ozone, and chlorine dioxide.

oxidation A chemical combination with oxygen. In a broader sense, the loss of one or more electrons (negative valence) from an element or a radical or an increase in positive valence.

oxidation–reduction A chemical reaction in which one reactant is reduced (gains one or more electrons) and another is oxidized (loses one or more electrons). The process based on this reaction is referred to commonly as redox. (*See redox.*)

pH A measure of the acidity or alkalinity of a solution, such that a value of 7 is neutral; lower numbers represent acidic solutions and higher numbers alkaline solutions. Strictly speaking, pH is the negative logarithm of the hydrogen ion concentration (in moles per liter). For example, if the concentration of hydrogen is 10^{-7} moles per liter, the pH will be 7.

pilot testing Evaluation, on a scale larger than laboratory scale but smaller than full scale, of the amenability of water to treatment by particular operations or processes.

potable water Water that is safe and satisfactory for drinking and cooking.

pretreatment One or more physical or chemical processes occurring immediately before water enters a downstream treatment unit process to enhance the downstream process efficiency. An example of pretreatment for ion exchange is the use of prefiltration to reduce suspended solids before an ion exchange system to prevent channeling.

raw water Water as it comes from the source (well, lake, reservoir, river) or untreated water.

regenerant A chemical solution used to restore an exhausted bed of ion-exchange resin to the fully ionic (regenerated) form necessary for the desired ion exchange to again take place effectively. Regenerant contains excess ions, which drive the reaction equilibrium to the desired form. For example, sodium chloride (brine) is often used as a regenerant.

regeneration The periodic restoration of an ion-exchange resin capacity back to a usable form by employing a regenerant to displace ions removed during the treatment process. Ion-exchange resins are regenerated by reversing the equilibrium reaction between the exchanging ions and the ions that are removed during treatment and retained on the resin. The regenerant typically contains a high concentration of the ex-changing ions to drive the equilibrium reaction in a reverse direction from the normal service cycle.

regeneration efficiency In the regeneration of ion-exchange resin, the degree to which regenerant ions are utilized on an equivalent basis to those ions replaced.

residual The amount remaining after a process has been completed. For example, a free chlorine residual of 0.5 mg/L is left over after all the other chlorine demands in a particular water have been satisfied. It can be calculated by the percent leakage immediately after regeneration to the influent ionic load. See also *regeneration.*

resin In the water-treatment industry, resin is often used as an abbreviation for ion-exchange resin. It is usually in the form of specifically manufactured organic polymer beads used in softening and other ion-exchange processes to remove dissolved ions from water. See also *ion exchanger.*

secondary maximum contaminant level (SMCL) A non-enforceable numerical limit set by USEPA or other regulatory agency for a contaminant on the basis of aesthetic effects to prevent an undesirable taste, odor, or appearance that could have an adverse impact on the public welfare

selective ion exchanger An ion-exchange medium that shows selectivity for particular contaminants. See also *selectivity*.

selectivity The relative capacity of an ion-exchange resin to remove different ions of similar charge and, therefore, the order in which specific ions are preferred by a given resin. The selectivity of an ion is influenced by valance or oxidation state and molecular weight.

soft water Having low hardness (low concentrations of calcium, magnesium, or both). There are no USEPA primary or secondary standards for water hardness. Water hardness as classified by the US Geological Survey uses the following classification:

Soft	0–60 mg/L as $CaCO_3$
Moderately Hard	61–120 mg/L as $CaCO_3$
Hard	121–180 mg/L as $CaCO_3$
Very Hard	>180 mg/L as $CaCO_3$

specific gravity The ratio of the density of a liquid substance to a standard density. For solids and liquids, the density is compared to the density of water.

strong acid An acid that is completely ionized in dilute water solutions. The quantitative measure of acid strength is the acid dissociation constant—i.e., the equilibrium constant for a chemical reaction of the following form (in which A represents an anion): $HA + H_2O = H_3O^+ + A^-$. For example, hydrochloric acid is completely ionized in water and is considered a strong acid.

strong acid cation (SAC) ion exchange resin A cation-exchange resin with an exchange site, i.e., an active group, usually sulfonic acid (RSO_3–H^+), that is ionized throughout most of the pH range (pH 1 to pH 4). Neutral salts (e.g., sodium chloride, NaCl; magnesium sulfate, $MgSO_4$; calcium nitrate, $Ca(NO_3)_2$) are said to be split by the resin to form their corresponding free acids (hydrochloric acid, HCl; sulfuric acid, H_2SO4; nitric acid, HNO_3).

strong base anion (SBA) ion exchange resin An anion-exchange resin with an exchange site, i.e., an active group, usually quaternary amine ($-N+(CH_3)_3$), that is ionized throughout most of the pH range (pH 0 to pH 13). Neutral salts (e.g., sodium chloride, NaCl; calcium sulfate, $CaSO_4$; potassium nitrate, KNO_3) are said to be split by the resin to form their corresponding free bases (sodium hydroxide, NaOH; calcium hydroxide, $Ca(OH)_2$; potassium hydroxide, KOH).

temporary hardness Hardness caused by the presence of dissolved carbonate and bicarbonate minerals (calcium bicarbonate and magnesium bicarbonate). Also known as carbonate hardness.

total dissolved solids (TDS) Minerals that dissolve in a liquid solution are called TDS. It can be measured as the weight per unit volume of solids remaining after a sample has been filtered through a 0.45 µm filter to remove suspended and colloidal solids. TDS is usually associated with ions that increase the conductivity of the water.

total organic carbon (TOC) A measure of the concentration of the organic carbon in water, determined by oxidation of the organic matter into carbon dioxide (CO_2). Total organic carbon includes all the carbon atoms covalently bonded in organic molecules.

total suspended solids (TSS) Solids that don't dissolve and can be filtered out of a liquid. To measure TSS, a well-mixed sample is filtered through a standard glass fiber filter, and the residue retained on the filter is dried to a constant weight at 217°F to 221°F. The increase in the weight of the filter represents the total suspended solids.

turbidity Cloudiness of water due to the scattering of visible light (the Tyndall effect) by particles such as clay, silt, and organic and inorganic matter. Most turbidity meters interpret turbidity as nephelometric turbidity units (NTU). Water with a high turbidity value has a cloudy or unclear appearance; turbidity of drinking water should not exceed 1.0 NTU.

uniformity coefficient A dimensionless factor to describe the uniformity of particle size in a filtration media. It is defined as the sieve size that passes 60 percent of the media grains by weight divided by the sieve size that passes 10 percent of the media grains by weight. Also shown as UC, and sometimes Uc. A UC of 1 would mean that all particles are exactly the same size (a situation not possible through commercial processes).

valence The number of electrons that an atom or radical can lose, gain, or share with other atoms or radicals. Valence means essentially combining power; it is the relative worth of an atom of an element in combining with the atoms of other elements to form compounds. The valence number of an element is the number of its electrons associated with formation of a particular compound. For example, the valence of Mn^{+6} is 6.

viscosity A term that describes a liquid's resistance to flow. Formally, it is the internal friction of a fluid caused by molecular attraction. The viscosity of water is influenced by temperature. Temperature affects the viscosity and density of water to such an extent that backwash rates routinely account for water temperature.

water softener A pressurized water treatment device in which hard water is passed through a bed of cation-exchange media (either inorganic or synthetic organic) for the purpose of exchanging calcium and magnesium ions for sodium or potassium ions, thus producing a softened water that is more desirable for laundering, bathing, and dishwashing. This cation-exchange process was originally called zeolite water softening. Most modern water softeners use a sulfonated bead form of styrene ($C_6H_5CH{:}CH_2$)/divinyl benzene ($C_6H_4(CH{:}CH_2)_2$) cation resin. See also *hard water*.

water softening The removal of calcium and magnesium ions, which are the principal causes of hardness in water. The cation-exchange resin method is most commonly used for residential and commercial water treatment.

weak acid An acid that is only partially ionized in dilute water solutions. The quantitative measure of acid strength is the acid dissociated constant. For example, acetic acid (CH_3COOH) is a weak acid. See also *acid*.

weak acid ion exchange (WAC) resin Those cation-exchange products with functional groups that are not capable of splitting neutral salts to form corresponding free acids. Weak acid cation-exchange resins have a much higher (three to four times higher) regeneration efficiency and capacity than their strong acid counterparts. They can only exchange ions in the neutral to alkaline pH range. Thus, they can remove carbonate hardness. They cannot remove noncarbonate hardness as the pH would drop too low.

weak base anion exchange (WBA) resin WBA, also known as acid adsorbers, do not have exchangeable ions until the free base sites on the resin are protonated by an acid. These resins adsorb free mineral acidity (Cl and SO_4). WBA resins have amine functional groups, and are typically regenerated with sodium hydroxide (NaOH), ammonia (NH_3), or sodium carbonate (Na_2CO_3).

zeolite Any of a large group of natural hydrous aluminum silicates of sodium, calcium, potassium, or barium, chiefly found in cavities of igneous rocks and characterized by a ready loss or gain of water by hydration; many are capable of ion exchange with solutions.

INDEX

D

E

F

G

H

I

O

P

R

S

T

U

W

Y